Lecture Notes in Earth Sciences 123

Editors:

J. Reitner, Göttingen
M. H. Trauth, Potsdam
K. Stüwe, Graz
D. Yuen, USA

Founding Editors:

G. M. Friedman, Brooklyn and Troy
A. Seilacher, Tübingen and Yale

For further volumes:
http://www.springer.com/series/772

Huilin Xing · Xiwei Xu

M8.0 Wenchuan Earthquake

Dr. Huilin Xing
University of Queensland
Earth Systems Science Computational
  Centre (ESSCC)
School of Earth Sciences
St. Lucia, Queensland 4072
Australia
h.xing@uq.edu.au

Dr. Xiwei Xu
China Earthquake Administration
Institute of Geology
Qijiahuozi, Deshengmenwai
Beijing 100029
People's Republic of China
xiweixu@vip.sina.com

ISSN 0930-0317
ISBN 978-3-642-01875-6   e-ISBN 978-3-642-01901-2
DOI 10.1007/978-3-642-01901-2
Springer Heidelberg Dordrecht London New York

Library of Congress Control Number: 2010938032

© Springer-Verlag Berlin Heidelberg 2011

This work is subject to copyright. All rights are reserved, whether the whole or part of the material is concerned, specifically the rights of translation, reprinting, reuse of illustrations, recitation, broadcasting, reproduction on microfilm or in any other way, and storage in data banks. Duplication of this publication or parts thereof is permitted only under the provisions of the German Copyright Law of September 9, 1965, in its current version, and permission for use must always be obtained from Springer. Violations are liable to prosecution under the German Copyright Law.

The use of general descriptive names, registered names, trademarks, etc. in this publication does not imply, even in the absence of a specific statement, that such names are exempt from the relevant protective laws and regulations and therefore free for general use.

Cover design: Integra Software Services Pvt. Ltd., Pondicherry

Printed on acid-free paper

Springer is part of Springer Science+Business Media (www.springer.com)

Preface

The great Wenchuan *Ms* 8.0 earthquake of 14:28 May 12, 2008 (called the Wenchuan earthquake hereafter) has shocked the world. It is the largest earthquake in the mainland of China in the past 60 years, causing the most serious damages, the largest stricken areas and the greatest difficulty for disaster relief. As determined by the China Earthquake Network Center, the focal depth of the Wenchuan earthquake is 19 km, and the earthquake occurred on the middle segment of the Longmenshan thrust belt along the southeast margin of the Qinghai-Tibetan Plateau. The event completely destroyed Yingxiu Town, Wenchuan County, Qushan Town, Beichuan County and Qingchuan County in Sichuan Province, while the stricken areas involved 48,810 villages and 4,667 towns for 417 counties (cities or districts) in 10 provinces (municipalities or autonomous regions) including Sichuan, Gansu, Shanxi and Chongqing, with a total area of about 500,000 km^2 and 46.25 million sufferers. According to the official report up to September 28, 2008, it is known that 69,227 people were killed, 374,643 injured and 17,823 missing during the earthquake, while about 15.1 million people needed to be urgently moved away and settled down. The earthquake has caused enormous damages of buildings and basic facilities, heavy losses of industrial and agricultural productions, and direct economic losses of 845.1 billions Yuan (RMB). Moreover, the subsidiary hazards induced by the earthquake, such as collapse, landslide, debris flows and earthquake lakes are of unprecedented all over the world.

The Sichuan-Yunnan faulted-block in southwestern China, where the Wenchuan earthquake occurred, is a part of the southeastward extrusive active tectonic system in the eastern and southeastern Qinghai-Tibetan plateau. A lot of strong earthquakes occurred historically in this region especially on the eastern boundary of this faulted block, and thus a few of major national and international projects were/are carried out to study the geological setting, fault systems and earthquake dynamics. For example, one of the author of this book, Xiwen Xu, has led the research activities on active fault system in this region supported by a few of national projects including the National Key Basic Research Projects (No.2004CB418401 & G1998040701) since 1999, which further led to an international collaboration commenced in 2006 between the two authors and their groups to investigate the fault system dynamics and its associated earthquakes in this region using the finite element code developed by Xing et al. at The University of Queensland. The Wenchuan earthquake occurred as a surprise, but more surprisingly to us is that it was located in a high risk zone of our preliminary simulation result done in May 2007, just 1 year before the Wenchuan earthquake occurred. This drives us to look deeply into the related issues.

Soon after the Wenchuan earthquake, under the organization of the China Earthquake Administration (CEA), the researchers from the Institute of Geology, China Earthquake Administration (IGCEA), went right to the epicentral areas to carry out emergency rescue and scientific investigation, and a large amount of data on the earthquake surface ruptures have been obtained at the first time. Moreover, after the accomplishment of the emergency rescue and urgent scientific investigation work, under the guidance of the State Wenchuan Earthquake Expert Committee, the researchers from IGCEA and the other related scientific institutions went again to the epicentral areas to carry out systematic scientific investigation, including the precise observation, measurement and record of the surface ruptures and earthquake hazards. Field observations show that the Wenchuan earthquake ruptured the two NW-trending imbricated reverse faults, the Beichuan-Yingxiu fault and the Guanxian-Jiangyou fault of the Longmenshan thrust belt. Among them, the earthquake intensity at Yingxiu, Longmenshan, Yuejiashan, Gaochuan, Chaping, Qushan and Nanba towns (villages) along the Beichuan-Yingxiu fault reaches up to XI, while the surface rupture zone is about 240 km long, dominated mainly by reverse faulting with right-slip component. The maximum vertical offset at the site to the north of Beichuan is 6.5±0.5 m, and the maximum right-lateral displacement is 4.9 m. The surface rupture zone to the south of Beichuan is dominated mainly by thrusting, where the maximum vertical offset is 6.2±0.5 m. The earthquake intensity at Bailu, Jinhua and Hanwang towns (villages) along the Guanxian-Jiangyou fault reaches X. The surface rupture zone is 72 km long and appears as a pure reverse faulting rupture with a maximum vertical offset of 3.5 m. In addition, a NW-trending surface rupture zone dominated by left-lateral strike slip faulting with vertical thrusting component is developed to the west of the aforementioned two surface rupture zones, having a length of about 7 km. The surface rupture pattern produced by the Wenchuan earthquake is the most complicated of recent great earthquakes and is the longest among the coseismic surface rupture zones for reverse faulting events ever reported in the intraplate settings. Inversions of seismic data have further indicated that the Wenchuan earthquake can further be resolved into two seismic faulting sub-events with 6~9 m coseismic offset that propagated to the northeast on a 300 km long, moderately-dipping (~33°) fault along the Longmenshan thrust fault zone. The sub-event nearby Yingxiu Town underwent oblique right-lateral thrusting slip, while the northeast sub-event nearby Beichuan (Qushan Town) exhibited primarily right-lateral displacement. The average offset on the focal fault plane reaches 5 m. The oblique-slip reverse fault type surface rupture associated with the Wenchuan earthquake indicates that the horizontal motion of the block in the middle and eastern part of the Qinghai-Tibetan Plateau has been transformed into crustal shortening and uplift along the Longmenshan thrust belt in between the South China and Bayan Har blocks.

This book is a production of our relevant research and field investigation as above. It includes 5 chapters describing the tectonic setting, historical earthquakes, the Wenchuan earthquake and aftershocks, numerical investigation of earthquake nucleation and occurrence, earthquake induced surface ruptures, disasters and damage features. The field observations of earthquake induced surface fractures and building damage, form a major and special part of this book and include a large number of digital photos with accompanying brief explanations. This collection of such photos in this book reflects authentically the characteristics of surface deformation and hazards produced by the Wenchuan earthquake. It may provide not only the objective historic record of this exceptionally great earthquake, but also basic data for earthquake research and engineering seismic design. We hope this book helpful for both earthquake scientists and the public in summing up of experience and lesson given by this earthquake for the countermeasure and disaster mitigation of large earthquake in the future, and further provide some new insights into the earthquake forecasting, distribution features of earthquake hazards occurred on steeply dipping oblique-slip thrust belt, as well as the tectonic movement and the uplift of the Qinghai-Tibetan plateau.

We would like to express our deep appreciation to Professor Yoko Ota of Yokohama National University, Professor Yue-Gau Chen of Taiwan University and Dr Y. Wang of CSIRO for constructive reviews, and to Professor Qidong Deng of Institute of Geology, China Earthquake Administration, Professor Yaolin Shi of Graduate University of Chinese Academy of Sciences, Professor Jiwen Teng of Institute of Geology and Geophysics, Chinese Academy of Sciences, Professor Mian Liu of Missouri University for helpful discussions, and to Dr Chris Bendall and Janet Sterritt-Brunner of Springer for their encouragement and help to have such a special topic on Wenchuan earthquake collected and published as a book.

Brisbane, Australia
Beijing, People's Republic of China
December 1, 2009

Huilin Xing
Xiwei Xu

Acknowledgements

The scientific investigations of the Wenchuan earthquake were carried out under the guidance and organization of Mr. Chen Jianmin, the director of CEA, Mr. Liu Yuchen and Mr. Xiu Jigang, the deputy directors of CEA, Mr. Zhao Heping and Mr. Li Ke, the heads of the Wenchuan Earthquake Commanding Headquarters, Mr. Wu Yueqiang and Miao Chonggang, the vice commanders of the Headquarters, and the other leadership of the Headquarters, Mr. Huang Jianfa and Mr. Lu Shoude, the department directors of CEA, Mr. Zhang Peizhen, the director of Institute of Geology, CEA, Mr. Li Ming, vice department director of CEA, and Mrs. Tian Liu, section chief of CEA, as well as the Emergency Scientific Investigation Team of Wenchuan earthquake and the State Wenchuan Earthquake Expert Committee. We thank also the Department of Earthquake Damage Protection, Department of Earthquake Emergency Response and Relief, Department of Personnel, Education, Science and Technology, and Department of Monitoring and Prediction of CEA, as well as Mrs. Jiang Zhao, the vice director of Institute of Geology, CEA for their enthusiastic supports. Many thanks also to Mr. Liu Benzhang, Shi Lei, Kang Darong, Liu Tao, Yu Yong, Zhang Degang, Luo Xinhua, Zhou Guanghua, Guo Anjun and Chu Xiaoyuan, the drivers from Earthquake Administration of Sichuan Province, as well as the volunteer drivers Mr. Xia Xiaolong, Lu Wei, Suo Yalun, Li Yuhui, Liu Xun, Wang Qing and Liu Dongguang from Beijing Offroader Association Club. The Mike Measuring Instrument Company is greatly appreciated for providing gratuitously field measurement instruments, and we are very grateful to the engineers form the company Mr. Liang Shipin, Luo Hancheng, Du Bin, Jin Min and Li Fushun for their selfless helps in technician training. Especially, we thank the people in the earthquake stricken areas for their helps in various aspects. They provided the Earthquake Emergency Scientific Investigation Team with indispensable technical conditions and spiritual supports.

This collection of photos in Chaps. 4 and 5 is mainly from the results of the scientific investigation of the Wenchuan earthquake. We thank our colleagues: Xueze Wen, Yongkang Ran, Jie Chen, Gongmin Yin, Honglin He, Zhihui Deng, Zhujun Han, Xiaoping Yang, Jianqing Ye, Baoqi Ma, Rongjun Zhou, Yulin He, Qinjian Tian, Guihua Yu, Shimin Zhang, Zhaomin Sun, Zhicai Wang, Xiguang Li, Zhuojun Chen, Tianyong Ge, Shen'e Yu, Youqing Ye, Yongbin Han, Yongjian Cai, Shiyuan

Wang, Haiqing Sun, Wentao Ma, Chenxia Li, Yanfen An, Shaopeng Dong, Zhanyu Wei, Feng Li, Zhongtai He, Yi Du, Shixue Jia, Jisheng Zhao, Youmei Wang, Guili Qi, for sharing the photos taken during the scientific investigation after the Wenchuan earthquake.

This work is supported by the Natural Science Foundation of China (grant No. 40821160550, 40974057 & 40728004), International Scientific joint project of China (grant No.2009DFA21280), Institute of Geology of CEA, Australian Research Council and The University of Queensland.

Contents

1 Tectonic Setting of Chuan-Dian Region 1
  1.1 Seismotectonic Setting 1
  1.2 Recent Crustal Movement 3

2 Historical Earthquakes, Ms8.0 Wenchuan Earthquake
  and Its Aftershocks 9
  2.1 Historical Earthquakes 9
  2.2 Wenchuan Earthquake 11
  2.3 Aftershocks 16

3 Earthquake Nucleation and Occurrence – Numerical Investigation . 21
  3.1 Introduction 21
  3.2 Tectonic Setting 22
    3.2.1 Xianshuihe Fault 23
    3.2.2 Anninghe and Daliangshan Fault 24
    3.2.3 Zemuhe and Xiaojiang Fault 24
    3.2.4 Longmenshan Thrust Belt 25
  3.3 Computational Method 25
  3.4 Numerical Analysis of Earthquake Activities 26
    3.4.1 Computational Model 26
    3.4.2 Simulation Results 29
  3.5 Conclusions and Discussions 37

4 Earthquake Surface Ruptures 39
  4.1 Introduction 39
  4.2 Beichuan Surface Rupture Zone 45
  4.3 Hanwang Surface Rupture Zone 102
  4.4 Xiaoyudong Surface Rupture Zone 126
  4.5 Appendix 4.1: Co-seismic Offsets Along the Beichuan
      Rupture Zone 139
  4.6 Appendix 4.2: Co-seismic Offsets Along the Hanwang
      Rupture Zone 146
  4.7 Appendix 4.3: Co-seismic Offsets Along the Xiaoyudong
      Rupture Zone 148

5 Earthquake Disasters and Damage Features of Structures and Buildings 149

    5.1   Seismic Intensity Distribution 149

    5.2   Niu Juan Gou Hypocenter of Earthquake 151

    5.3   Yingxiu Town 153

    5.4   Qushan Town 154

    5.5   Road and Bridge Damages 157

    5.6   Other Damages 162

    5.7   What Learned from Wenchuan Earthquake 167

References 183

Index 189

Chapter 1
Tectonic Setting of Chuan-Dian Region

1.1 Seismotectonic Setting

The NEE-directing motion of the Indian plate and its collision into the Eurasian plate has resulted in the formation of the world roof – the Qinghai-Tibetan Plateau with an elevation of over 4,500 m. The crustal thickness of the plateau reaches up to 60~70 km, about 20~30 km thicker than that of the Sichuan basin, where the crustal thickness is only about 40 km. Numerous large-scale strike-slip faults have been developed within the plateau. Among them, the right-lateral strike-slip faults include the NW-trending Kara Korum fault zone in the west and the nearly EW-trending Jiali fault, while the left-lateral strike-slip faults include the NE-trending Altyn Tagh, the Haiyuan, the Eastern Kunlun and the Garze-Yushu and the Xianshuihe faults. They dissect the Qinghai-Tibetan Plateau into the Lasha, Qiangtang, Bayan Har, Qaidam-Qilianshan, Myitkyina-Ximeng and Sichuan-Yunnan blocks. At the eastern margin of the Qinghai-Tibetan Plateau, there are many thrust belts resulted from the strong compression between these blocks and the Alasan, Ordos and South China blocks in the east (Fig. 1.1). The Wenchuan earthquake occurred on the middle segment of the Longmenshan thrust belt of the eastern Qinghai-Tibetan Plateau, where the Bayan Har block interacted strongly with the South China block (Fig. 1.2).

The Longmenshan thrust belt is located between the Bayan Har and South China blocks at the eastern margin of the Qinghai-Tibetan Plateau, making up the southeastern boundary of the Bayan Har block. It connects with the Xianshuihe and Anninghe faults in the west, borders on the Sichuan Basin in the south and Longmenshan mountain area in the north, and connects with the eastern margin of the Qinling Mountains in the east. It is an important part of the North-South seismo-tectonic belt on the mainland of China. This seismitectonic belt, together with the NW-trending eastern Kunlun fault, nearly EW-trending Minjiang uplift zone, and NW-trending Garze-Yushu and Xianshuihe faults, makes up the tectonic boundary of the Bayan Har block (Fig. 1.2).

The Longmenshan thrust belt spanning about 500 km long and 40~50 km wide consists of the Wenchuan-Maowen fault (F1), Beichuan-Yingxiu fault (F2), Pengguan fault (F3) and a piedmont buried fault under the Chengdu plain (Fig. 1.2). All of them are dominated by right-lateral strike-slip with thrusting component. The northern segment of the Longmenshan thrust belt comprises early to middle

H. Xing, X. Xu, *M8.0 Wenchuan Earthquake*, Lecture Notes in Earth Sciences 123, DOI 10.1007/978-3-642-01901-2_1, © Springer-Verlag Berlin Heidelberg 2011

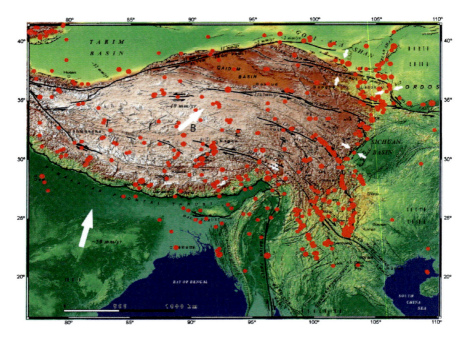

Fig. 1.1 Sketch map showing the block motion and tectonic transformation at the eastern margin of the Qinghai-Tibetan Plateau and its adjacent regions (Tapponnier et al., 2001; Xu et al., 2006; Xu 2009)

Pleistocene faults, while the middle segment Holocene faults, and the southern segment have been active since late Quaternary. Since late Quaternary, the right-lateral slip rate of the active faults in the Longmenshan thrust belt is about 1 mm/year (Ma et al., 2005) or 1~10 mm/year (Densmore et al., 2007), while the vertical slip rate of thrusting is ≤ 1 mm/year (Ma et al., 2005; Tang and Han, 1993; Densmore et al., 2007). However, some researchers have also pointed out that the vertical slip rate of the Beichuan-Yingxiu fault on the middle segment of the Longmenshan thrust belt reaches 1–2 mm/year, and postulated that the general slip rate of the Longmenshan thrust belt is 4~6 mm/year (Deng et al., 1994, 2002; Zhao et al., 1994). The NW-directing crustal shortening rate of this thrust belt as deduced from geologic data is 10 mm/year (Avouac and Tapponnier, 1993), while the crustal shortening ratio as determined from balance profile is 40%~60% (Lin and Wu, 1991).

Prior to the occurrence of the Wenchuan earthquake in 2008, in addition to the occurrence of a M(magnitude) 7.5 earthquake at Diexi on the Minshan uplift in the vicinity of Songpan and an earthquake swarm consisting of three M6.7~7.2 events at Songpan, the following earthquakes occurred around the Longmenshan region: an earthquake of M6.5 occurred at Mianchi-Caopo of Wenchuan County along the middle segment of the Wenchuan-Maowen fault in 1657, an earthquake of M6.2 occurred at Chaping, Beichuan County along the middle section of the Beichuan-Yingxiu fault in 1958, and three M6~6.2 events occurred along the

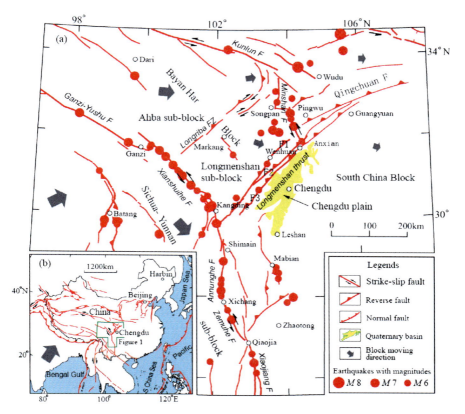

Fig. 1.2 Seismotectonic map of the Wenchuan earthquake with major historical earthquakes till 2008 (Xu et al., 2008b)

southern segment in 1327, 1828 and 1970 (Jones et al., 1984; Chen et al., 1994; Department of Earthquake Hazard Prevention, China Earthquake Administration, 1999). The Wenchuan earthquake is the largest earthquake among those that have occurred in the Longmenshan region since the beginning of historical records, and it is also a destructive earthquake that has caused the most serious economic loses and greatest difficulties in disaster relief. The earthquake is considered to be a continental earthquake resulted from the simultaneous rupturing of multiple imbricate reverse faults.

1.2 Recent Crustal Movement

The mainland of China is located at a critical tectonic position among the convergence of the Eurasian, Indian and Philippine Sea plates, where large-scale crustal deformation occurs. The GPS velocities inferred in Eurasian-fixed frame shows that the maximum velocity reaches up to 40–42 mm/year at the foreland of the Himalayas in N20°E-direction. Northward it becomes smaller to ~19 mm/year at Qaidam Basin, and only 10~12 mm/year at Junggar Basin. The velocity loss occurs in the Tianshan to transform to recent uplift on widespread active faults and folds, and to attest to a rapid crustal shortening, with a crustal shortening rate of ~20 mm/year for the western Tianshan to the west of Kashi, ~13 mm/year for the middle Tianshan and less than 2 mm/year for the eastern Tianshan. The crustal velocity at Altai Region is only several mm/year. The crustal movement turns to eastward at the middle and eastern Qinghai-Tibetan Plateau, and it becomes NE at the Hexi Corridor on the northeast corner of the plateau, NEE to nearly EW at the junction areas of Gansu, Sichuan and Qinghai Provinces and the Longmenshan region, SEE at the western Sichuan, SSE at the middle Yunnan province and even to SW at the southwest Yunnan province. All these may indicate that clockwise rotation of the middle and eastern Qinghai-Tibetan Plateau and the Sichuan-Yunnan region occurs around the eastern Himalayan syntaxis. At the same time, the eastern margin of the plateau has absorbed 4~6 mm/year velocity to be transformed into the uplift of the plateau itself. The Northeast China is a relatively stable region, where the velocity as monitored by the GPS observation points is less than 4 mm/year with relatively poor consistency. The South China and North China blocks as a whole tend to move to SEE relative to the stable Eurasian plate (Europe and Siberia), and the velocity vector becomes greater southward with a value between 8 and 12 mm/year along the southeast coast (Fig. 1.3).

Recently, GPS observations showed that the modern crustal shortening across the Longmenshan thrust belt was insignificant (King et al., 1997; Lu et al., 2003; Shen et al., 2005). The modern crustal shortening rate within a width of 700 km across the Longmenshan thrust belt is ~7 mm/year (Wang et al., 2002), or less than 4 mm/year within a width of 250 km (Zhang et al., 2004), with a left-lateral slip rate of 7.5 mm/year (Zhang et al., 2003). According to limited geologic and GPS data across the Longmenshan region, the modern crustal shortening rate has been postulated to be less than 3 mm/year (Shen et al., 2005; Meade, 2007; Densmore et al., 2007; Burchfiel et al., 2008). Although no measured deformation data over time scales longer than the recurrence interval between two progressive events comparable to the Wenchuan earthquake is available so far, the shortening rate of about 3 mm/year as obtained from GPS monitoring data for more than 10 years prior to the Wenchuan earthquake (Zhang et al., 1991; Shen et al., 2005; Meade, 2007) indicates to a great extent that the Longmenshan thrust belt has been strongly locked before the earthquake, rather than to represent the long term crustal shortening rate (Xu et al., 2008a, b) (Fig. 1.4).

The data obtained from 50 CGPS observation points, two periods of mobile observations at 100 GPS campaigns carried out before and after the earthquake and 145 GPS post-earthquake emergency campaigns in the Wenchuan earthquake epicenter and its adjacent regions have provided us with near-fault coseismic horizontal slip field associated with the Wenchuan earthquake (Fig. 1.5). The observation points with greater horizontal slips are located in Sichuan Basin and the Longmenshan region near the surface rupture zones, while the slip decreases significantly in the other areas, indicating the feature of deformation localization.

1.2 Recent Crustal Movement

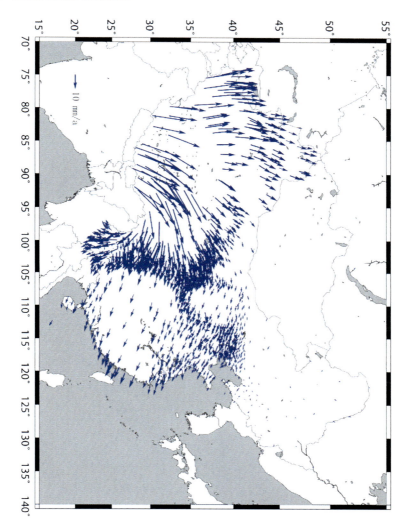

Fig. 1.3 GPS velocities (mm/year) in and around the mainland of China with respect to stable Eurasia for the period of 1999–2007 (by courtesy of Wang Min)

Among them, the near fault GPS points on the foot-wall of the surface rupture zone (in the Sichuan Basin side) show consistently a northwestward motion with slips of 1~1.5 m, and the maximum slip at observation point H035 may reach up to 2.43 m. On the hanging-wall (Longmenshan side), the observation points are located relatively far away from the surface rupture zone, so that the near fault slip can not be obtained, and the SE-directing slip at the adjacent points is less than 1 m in general. The GPS-monitored northwestward opposite motion of the Sichuan Basin (the South China block) during the earthquake should reflect an elastic rebound of the crust.

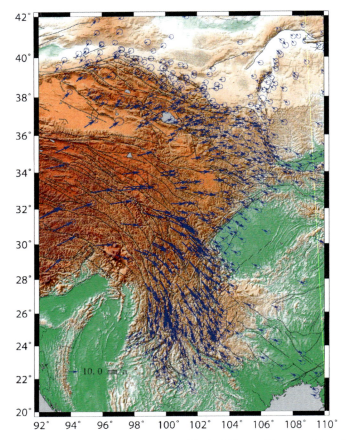

Fig. 1.4 GPS velocities (mm/year) of the eastern Qinghai-Tibetan Plateau and its adjacent areas with respect to stable Eurasia for the period of 1999–2007 (by courtesy of Wang Min)

Pre-seismic and postseismic leveling data indicate that the western side of the Beichuan-Yingxiu rupture zone mainly presented coseismic uplift with respect to the reference point at Pingwu county seat, and the maximum uplift was observed to be 4.71 m at Qushan Town, Beichuan County, about 100 m west of the fault scarp. However, its eastern side or foot-wall showed a subsidence with maximum value of 0.6 m near the rupture between Qushan Town and Guixi Town, then this value descended for 0.3~0.4 m, and only about 4 cm at Jiangyou. The leveling points northeast of Qushan Town are distributed along the fault valley, and the distance of all the points to the seimogenic fault is generally less than 500 m. According to the leveling results, it can be determined that the maximum coseismic vertical offset associated with the Wenchuan earthquake at sites from Qushan Town to Guixi Town may reach up to 5.4 m, among which the absolute ascending amount on the western side of the fault is significantly greater than the descending amount on the eastern wall, and their ratio is 8:1.

1.2 Recent Crustal Movement

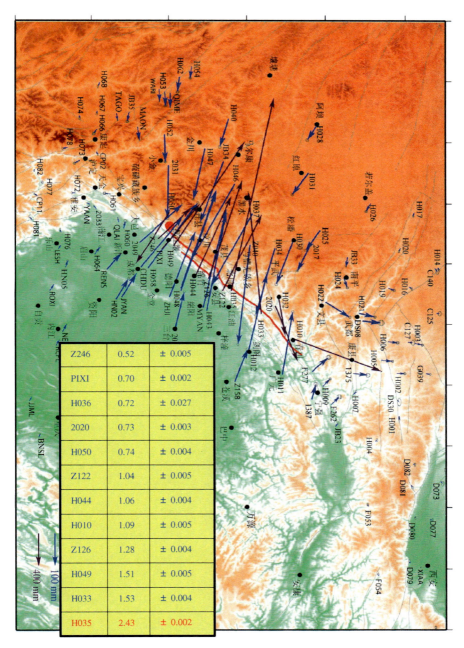

Fig. 1.5 GPS monitored co-seismic horizontal slip field of the Wenchuan earthquake. The inserted table shows a part of the observation point number, slip amount in meter and error (after the research group of crustal movement observation network of China, 2008)

Chapter 2
Historical Earthquakes, Ms8.0 Wenchuan Earthquake and Its Aftershocks

2.1 Historical Earthquakes

Current earthquake prediction methods were so far mostly based on recorded historical earthquakes. However, our knowledge of long-term earthquake behaviour is strongly restricted by the rarity of large earthquakes and the short historical and instrumental records of earthquakes, and hence our understanding could be misled by incomplete and unreliable data as well as uncertain analysis. These facts can certainly affect the progress of scientific research on seismic hazard assessment and earthquake forecast. Therefore, for a seismogenic fault, it is important to determine a complete and reliable long-term history of earthquake ruptures on all segments of the fault if possible. Most studies on this topic attempted to determine rupture histories from distributions of damage, surface ruptures, and aftershocks of historical and modern earthquakes.

The Longmenshan sub-block among the Sichuan-Yunnan rhombic, the Bayan Har and South China blocks (Figs. 2.1 and 2.2; Xu et al., 2003) in southwestern China, where the Wenchuan earthquake occurred, is a part of the southeastward extrusive active tectonic system in the southeastern Qinghai-Tibet plateau (Tapponnier et al., 1982). A lot of strong earthquakes occurred in this region historically as shown in Table 2.1 and Fig. 2.1. Especially for the boundary fault system between the Sichuan-Yunnan rhombic block and the Bayan Har block or the South China block, which consists of four major faults, the Xianshuihe, Anninghe, Zemuhe and Xiaojiang faults, as well as other secondary faults, e.g. the Daliangshan fault, forms a huge left-lateral strike-slip fault system with a total length of over 1,100 km (Fig. 2.1). This was once named the "Kangting fault system" (Tapponnier and Molnar, 1977). The strong tectonic and seismic activity and relatively long (several hundreds of years) record of earthquakes (Allen et al., 1991; Wen, 1993, 2000) make this fault system a promising site to study rupturing history and behaviour of earthquakes on active strike-slip fault zones in the intraplate region of China (Fig. 2.1). Besides of those earthquakes occurred along the above Kangting fault system, around Longmenshan thrust belt, a few of medium and large earthquakes occurred before Wenchuan earthquake, such as a magnitude 7.5 earthquake at Diexi on the Minshan uplift in the vicinity of Songpan and an

H. Xing, X. Xu, *M8.0 Wenchuan Earthquake*, Lecture Notes in Earth Sciences 123, DOI 10.1007/978-3-642-01901-2_2, © Springer-Verlag Berlin Heidelberg 2011

2 Historical Earthquakes, Ms8.0 Wenchuan Earthquake and Its Aftershocks

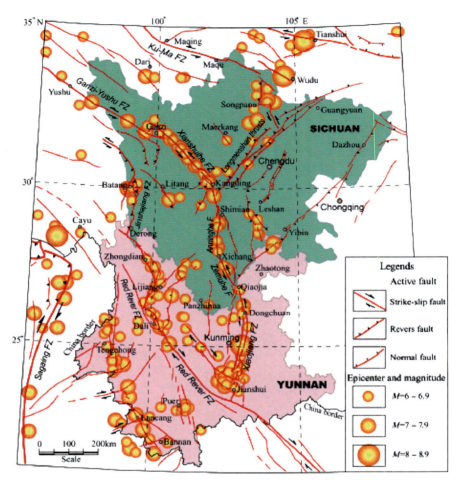

Fig. 2.1 Major historical earthquakes recorded in the Sichuan and Yunnan region

earthquake swarm consisting of three M6.7~7.2 events at Songpan, an earthquake of magnitude 6.5 occurred at Mianchi-Caopo of Wenchuan County on the middle segment of the Wenchuan-Maowen fault of the thrust belt in 1657, an earthquake of magnitude 6.2 occurred at Chaping of the Beichuan County on the middle segment of the Beichuan-Yingxiu fault in 1958, and three M6.0~6.2 events occurred on the southern segment of the Longmenshan thrust belt in 1327, 1828 and 1970 (Jones et al., 1984; Chen et al., 1994; Department of Earthquake Hazard Prevention, China Earthquake Administration, 1999). Recently, a lot of microseismicity also occurred before the Wenchuan earthquake. Figure 2.2 shows the hypoDD relocated earthquakes from 1992 to 2008 (Zhu et al., 2008). However, there are no large earthquakes (>= M7.0) recorded along the Longmenshan thrust belt before the 2008 Wenchuan earthquake (Fig. 2.1). The Wenchuan earthquake occurred as a big surprise.

2.2 Wenchuan Earthquake

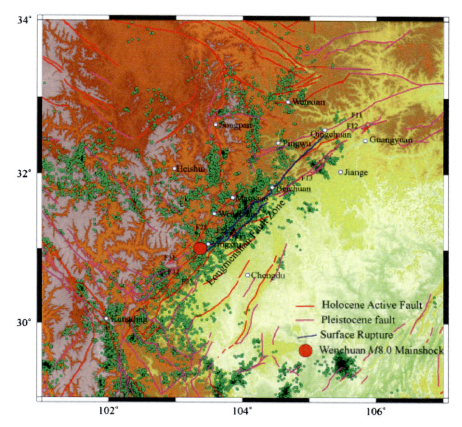

Fig. 2.2 The hypoDD relocated microseismicity before the Wenchuan earthquake from 1992 to 2008

Using data from historical and modern earthquakes and relevant geological investigations, the positions and spatial extents of ruptures of such major earthquakes could be achievable, such as the major earthquakes that occurred during the latest several 100 years along the boundary fault system between the Sichuan-Yunnan rhombic block and the Bayan Har block or the South China block are well studied to build a model of the historical rupture pattern (Fig. 2.3) and a further spatial-temporal pattern of rupture history (Fig. 2.4) (Wen et al., 2008)

2.2 Wenchuan Earthquake

Wenchuan earthquake occurred along the Longmenshan thrust belt as detailed in Table 2.2.

The 3-component acceleration waveforms recorded by strong motion seismograph in Longmenshan region (Figs. 2.5, 2.6 and 2.7) show that the peak

Table 2.1 List of major historical earthquakes in the Sichuan and Yunnan region

Event no.	Location				Seismic intensity distribution
	Year.month.day	Mag.	Lat.N	Lon.E	Ih
1	1216.03.24	7.0	28.4	103.8	IX
2	1515.06.27	7.8	26.7	100.7	X
3	1536.03.29	7.5	28.1	102.2	X
4	1652.07.13	7.0	25.2	100.6	IX+
5	1713.09.04	7.0	32.0	103.7	IX
6	1725.08.01	7.0	30.0	101.9	IX
7	1733.08.02	7.8	26.3	103.1	X
8	1786.06.01	7.8	29.9	102.0	$\geq$X
9	1789.06.07	7.0	24.2	102.9	IX+
10	1799.08.27	7.0	23.8	102.4	IX
11	1816.12.08	7.5	31.4	100.7	X
12	1833.09.06	8.0	25.0	103.0	$\geq$X
13	1850.09.12	7.5	27.7	102.4	X
14	1870.04.11	7.3	30.0	99.1	X
15	1879.07.01	8.0	33.2	104.7	XI
16	1887.12.16	7.0	23.7	102.5	IX+
17	1893.08.29	7.0	30.6	101.5	IX
18	1896.03	7.0	32.5	98.0	IX
19	1904.08.30	7.0	31.0	101.1	IX
20	1913.12.21	7.0	24.2	102.5	IX
21	1923.03.24	7.3	31.3	100.8	X
22	1925.03.16	7.0	25.7	100.2	IX+
23	1933.08.25	7.5	32.0	103.7	X
24	1941.05.16	7.0	23.7	99.4	IX
25	1941.12.26	7.0	22.1	100.0	VIII
26	1947.03.17	7.7	33.3	99.5	
27	1948.05.25	7.3	29.5	100.5	X
28	1950.08.15	8.6	28.5	96.0	>X
29	1955.04.14	7.5	30.0	101.8	X
30	1970.01.05	7.8	24.2	102.7	X+
31	1973.02.06	7.6	31.5	100.5	X
32	1974.05.11	7.1	28.2	104.1	IX
33	1976.05.29	7.4	24.6	98.7	IX
34	1976.05.29	7.3	24.5	99.0	IX
35	1976.08.16	7.2	32.6	104.1	IX
36	1976.08.23	7.2	32.5	104.3	VIII+
37	1988.11.06	7.2	23.2	99.6	X
38	1988.11.06	7.4	22.9	99.8	IX

Note: Here Ih is seismic intensity distribution (Chen et al., 1999).

acceleration on the seismogenic fault, earthquake surface rupture zones and their adjacent areas is relatively large, with the maximum value approximates to 1_g. The damages of buildings and constructions caused directly by co-seismic surface offset along the seismogenic fault, the long duration time of earthquake rupturing,

2.2 Wenchuan Earthquake

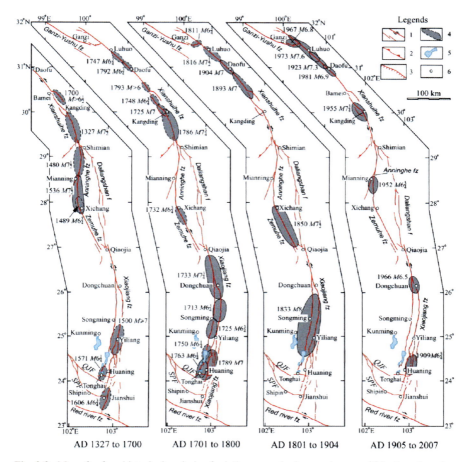

Fig. 2.3 Maps for four historical periods of relative severely damaged areas of 36>6 earthquakes boundary fault system between the Sichuan-Yunnan rhombic block and the Bayan Har block or the South China block. Legends: *(1)* Major (*thick line*) and secondary (*thin line*) active strike-slip faults; *(2)* active reverse fault; *(3)* active normal fault; *(4)* relative severely damaged areas of earthquakes with date and magnitude indicated, and *dashed lines* showing uncertain or inferred borders; *(5)* lakes; *(6)* cities or towns. Abbreviation for fault's name: QJF, Qujiang fault; SPF, Shiping fault (Wen et al., 2008)

secondary hazards such as large scale landslide, collapse and debris flow induced by high peak value acceleration, as well as the low level seismic design for building construction, are the main factors that caused the Wenchuan earthquake to produce exceptionally tremendous disasters.

The above waveforms recoded by strong motion seismograph show also that the Wenchuan earthquake consists of two sub-events (Figs. 2.6 and 2.7). The similar conclusion was also drawn from the inversion of seismic inversion (e.g. Fig. 2.8)

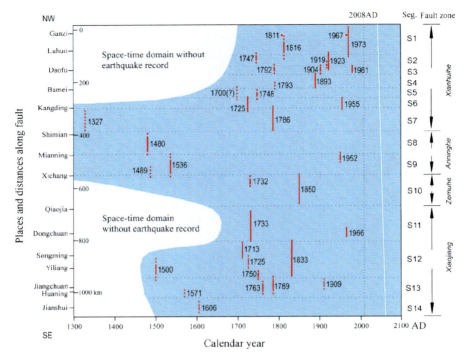

Fig. 2.4 A spatial-temporal pattern of the rupture history of *M* > 6 earthquakes along the boundary fault system between the Sichuan-Yunnan rhombic block and the Bayan Har block or the South China block. *Thick vertical lines* or *dashed lines* are projections of earthquake ruptures, with various classes of reliability for the determination of the ruptures: class A, *Solid lines*; class B, *broken lines* at two ends; and class C, *dashed lines*. Round dots are projections of those ruptures along a normal fault trending perpendicular to the Xianshuihe fault zone. Numerals show rupturing time in calendar years. Horizontal dot-lines show rupture segmentation. Fault names and segment symbols (S1–S14) are labeled on the left side of this diagram (Wen et al., 2008)

Table 2.2 Earthquake focal parameters

Magnitude	Moment magnitude: $M_w = 7.9$; Richter magnitude: $M_s = 8.0$
Occurrence time	14:28 LT, May 12, 2008
Initial rupturing point	31.1°N, 103.3°E
Focal depth	~19 km (initiation rupture point)
Occurrence site	Central Sichuan Province, China
Reference source	China Earthquake Network Center, U. S. Geological Survey, U. S. South California Seismic Center, Liu Qiyuan et al. (2008)

2.2 Wenchuan Earthquake

Fig. 2.5 3-component waveforms of the Wenchuan earthquake recorded by Western Sichuan seismograph array (after China earthquake network center)

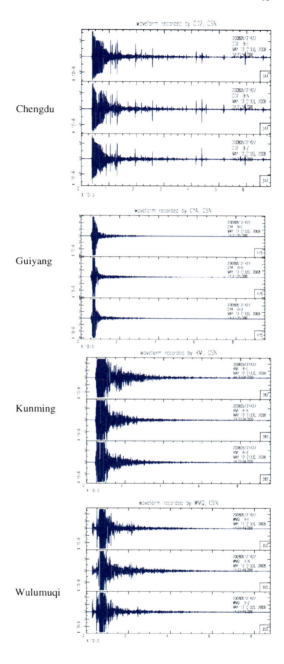

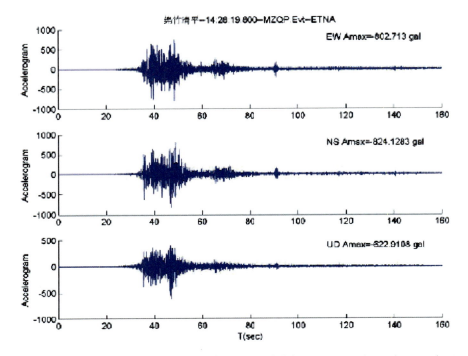

Fig. 2.6 3-component acceleration waveforms recorded by strong motion seismograph at Qingping, Mianzhu town on the central fault of the middle segment of the Longmenshan thrust belt (after China strong earthquake network center)

2.3 Aftershocks

As determined by the China Earthquake Network Center, the aftershocks of Wenchuan earthquake reaches up to 30,000, among which the largest one has a magnitude of 6.4 (Fig. 2.9). Zhu et al. (2008), Huang et al. (2008) and Chen et al. (2009a) have relocated the main shock and 3,622 aftershocks by adopting two different velocity structure models for both the northwest and southeast sides of the Longmenshan thrust belt and by using the double-difference earthquake location algorithm (Fig. 2.10). The vertical and horizontal errors in the relocation are 0.85 km and 0.75 km, respectively, while the root mean square residual error is 0.20 s. The relocation result indicates that the aftershocks are mostly concentrated along a NE-NW-directing zone on the northwest wall of the Beichuan-Yingxiu fault of the Longmenshan thrust belt with a total length of about 330 km, while the focal depth is within the range of 4~20 km. It should be pointed out, however, that the concentration zone of aftershocks is characterized significantly by segmentation. Taking the 60 km-long segment (from Beichuan to Nanba of Pingwu County, 104.5°E~104.8°E) with sparsely distributed aftershocks as a mark, we may divide the zone into 3 basic segments as (Figs. 2.2, 2.9, 2.10 and 4.1): the

2.3 Aftershocks

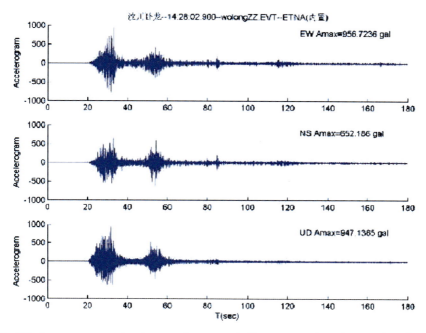

Fig. 2.7 3-component acceleration waveforms recorded by strong motion seismograph at Gengda county on the Back Range fault of the middle segment of the Longmenshan thrust belt (after China strong earthquake network center)

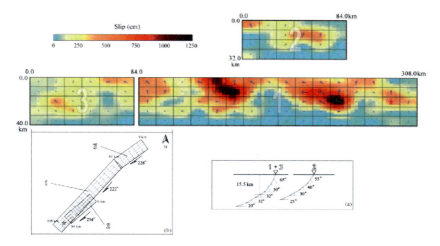

Fig. 2.8 The result of the inversion of seismic data modeling the segmented rupturing of two imbricate reverse faults shows that the Wenchuan earthquake can further be resolved into two seismic faulting sub-events with 6~9 m coseismic offset on a 33°-dipping strike-slip reverse fault and propagating northeastward. Among them, the sub-event occurred near the Yingxiu town was reverse faulting event with right-slip component, while that occurred near the Beichuan town was a right-lateral strike-slip faulting event with thrust component. The average offset on the seismic fault reaches up to 5 m (Parsons et al., 2008; Wang et al., 2008)

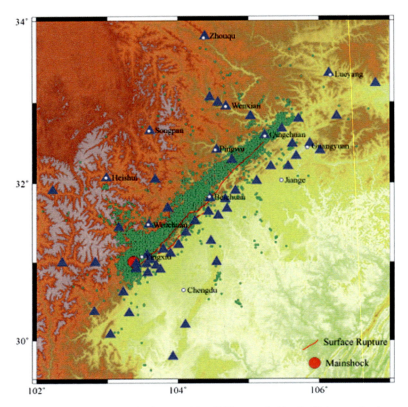

Fig. 2.9 Distribution of the aftershocks up to May 2009, recorded by the Sichuan seismic network. The blue triangular shows the stations used for aftershocks' relocation

Yingxiu-Beichuan, Beichuan-Nanba and Qingchuan segments. On the Yingxiu-Beichuan segment, the aftershocks are controlled mostly by the Beichuan-Yingxiu fault, and partly controlled by the Pengxian-Guanxian fault. This may indicate that the two imbricate reverse faults are characterized by steeply-dipping at the upper part and gently-dipping at the lower part, and they finally merge into a decollement at about 20 km depth, dominated mainly by reverse faulting with a small strike-slip component. In the vicinity of Xiaoyudong village, there is a 30 km-long, NW-SE-trending concentration zone of aftershocks, and it is postulated to be a secondary tear fault, dominated by left-lateral strike-slip. On the Beichuan-Nanba segment, the aftershocks are sparsely distributed mainly on the hanging wall (northwest side) of the Beichuan-Yingxiu fault, dominated equally by both right-slip and thrusting. The Qingchuan segment is the most peculiar concentration zone of aftershocks, appearing as a NE-trending zone, deviating significantly from the fault of the Longmenshan thrust belt, cutting obliquely the NEE-trending Qingchuan fault.

2.3 Aftershocks

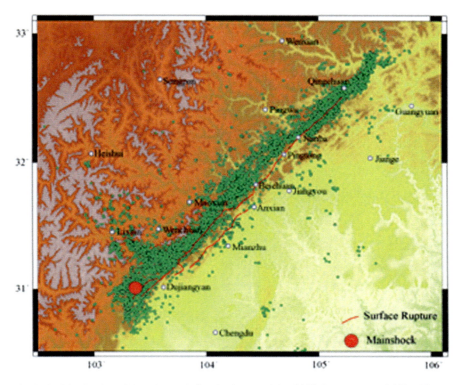

Fig. 2.10 Distribution of the relocated aftershocks up to May 2009 (by courtesy of Ailau Zhu)

Chapter 3
Earthquake Nucleation and Occurrence – Numerical Investigation

3.1 Introduction

The great Wenchuan earthquake caused much damage in Sichuan and neighbor provinces in Western China. As described above, it occurred along the Longmenshan thrust belt that is situated at the transformation site between the NEE-directing compressive stress field of the Qinghai-Tibetan plateau and the NW-directing compressive stress field of the western part of the South China block. Around the Longmenshan thrust belt is a complicated fault system (i.e. Chuan-Dian fault system) that collectively accommodate lithosphere motion in the eastern Tibetan Plateau (Fig. 3.1). The Chuan-Dian fault system in Southwestern China, containing several major faults including Ganzi-Yushu, Xianshuihe, Anninghe, Zemuhe, Xiaojiang, Daliangshan, Longmengshan, Honghe, Yulongxi and Lijiang-Xiaojinhe faults, is quite complicated and active, numerous major earthquakes have frequently occurred historically as detailed in Chap. 2. The former six ones form the eastern boundary of the Chuan-Dian rhombic block in southwestern China (It is simply denoted as the eastern boundary in this chapter if without special notation). It may be the place where the most frequent major earthquakes in the mainland of China occur, six large earthquakes have occurred along the Xianshuihe fault since 1893. However no major earthquakes were recorded along the Longmenshan thrust belt except few moderate earthquakes at its southwestern end before the Wenchuan earthquake (Figs. 2.1 and 3.2). That may be why most of the earthquake researches in this region were focused on the Xianshuihe fault and its adjoining major faults at the eastern boundary before the Wenchuan earthquake (Xu et al., 2003), while research focuses have been changed to Longmenshan thrust belt itself afterwards (Densmore et al., 2007; Xu et al., 2008a). Here we would like to take the Chuan-Dian fault system as a whole to investigate the earthquake activities to figure out the nucleation and occurrence of the Wenchuan earthquake and to further evaluate the future earthquake tendency.

An adaptive static/dynamic finite element based computational model and software has been developed (Xing and Makinouchi 2002a,b,c; Xing et al 2004, 2006, 2007a, 2007b and 2009) and applied here to investigate the earthquake nucleation and rupture behavior in the Chuan-Dian fault system referring the available

H. Xing, X. Xu, *M8.0 Wenchuan Earthquake*, Lecture Notes in Earth Sciences 123, DOI 10.1007/978-3-642-01901-2_3, © Springer-Verlag Berlin Heidelberg 2011

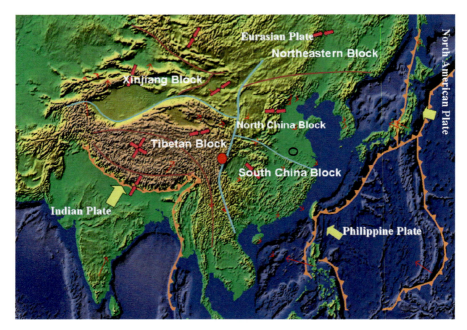

Fig. 3.1 Sketch map showing the regional dynamics of the mainland of China and its adjacent regions (Complied according to the data from Xie et al.)

information on fault geometry and long-term slip rate, as well as GPS data and historical strong earthquakes' occurrence. The data from the recorded earthquakes and relevant geological observations is also applied to help in numerically investigating this whole fault system towards an improved understanding the special earthquake activities in this region, such as why more earthquakes occurred along the Xianshuihe fault than the other faults, why the giant Ms8.0 Wenchuan earthquake occurred along the Longmenshan thrust belt after such a relatively long quiescent period in seismicity and how the earthquake ruptures interact with each other, and to further forecast the high risk fault segment in this region.

3.2 Tectonic Setting

To study earthquake activities in Chuan-Dian fault system, the following 10 major faults have been included (Fig. 3.2): Xianshuihe, Anninghe, Zemuhe, Xiaojiang, Daliangshan, Longmengshan, Honghe, Yulongxi and Lijiang-Xiaojinhe faults. The five major fault zones named Xianshuihe, Anninghe, Zemuhe, Xiaojiang and Daliangshan behave as a huge left lateral strike-slip active fault system form the eastern boundary of Chuan-Dian rhombic block in southwestern China, where the most frequent seismicity in the mainland of China occurs, and thus detailed below together with Longmenshan thrust belt.

3.2 Tectonic Setting   23

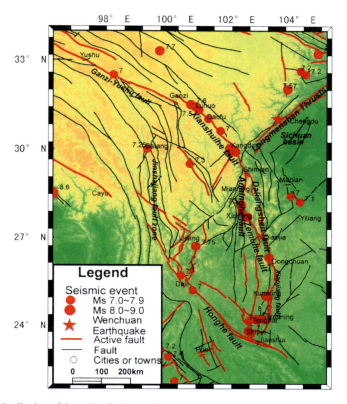

Fig. 3.2 Distribution of the major faults and historical earthquakes in the Chuan-Dian fault system, southeast margin of the Qinghai-Tibetan Plateau

3.2.1 Xianshuihe Fault

The Xianshuihe fault strikes N40°W, bending in the acuate form, slightly convex northeastward, for about 400 km in length. There exists a pull-apart, named the Garzê basin 22~25 km wide in the left step between the Garzê-Yushu fault and Xianshuihe fault. The northwestern segment of the Xianshuihe fault extends from Garzê to Bamei Town and consists of only a single fault. A large number of left-laterally offset landforms have developed and of them are relics of surface ruptures resulting from recent large earthquakes. The recent field investigation shows that the left lateral slip rate of the northwestern segment of the Xianshuihe fault reaches 14 mm/a and reduces to 10 mm/a for its southeastern segment, the Moxi fault. This slip rate loss is owing to the strike deflecting 15 degree from NW on its northwestern segment to NS on its southeastern segment, where the Moxi fault approaches to the NS trending Anninghe fault, to obstruct southeastern sliding of the block west of the fault (Xu et al., 2003, 2008a).

3.2.2 Anninghe and Daliangshan Fault

The northern segment of the eastern boundary of the middle Yunnan sub-block consists of the NS-trending Anninghe and Daliangshan faults (Fig. 3.2). Both faults converge at the southern end of the Xianshuihe fault near Shimian County, Sichuan Province. Toward south they are gradually separated and then converge again around Qiaojia County, Yunnan Province, to connect with the Xiaojiang fault to the south. The Anninghe fault is dominated by left lateral slip. It was reported that its left-lateral slip rate is in a range of 5.5–8.5 mm/a, the recent observation shows that the left-lateral slip rate of the southern segment of the Anninghe fault is only 6.5 mm/a. The Daliangshan fault lies in the hinterland of Daliang Mountain and strikes to 330°–360°. It consists of five *en echelon* secondary faults, from north to south, namely the Anshunchang-Xinchang fault, Shimian-Haitang-Yuexi fault, Puxiong-Zhuhe fault, Tuodu-Butuo fault, and Jifulala-Jiaojihe fault. Its left-lateral slip rate is 3.3 ± 0.3 mm/a. Therefore, the left-lateral slip rate of the northern segment of the eastern boundary of the middle Yunnan sub-block should be the sum of the rates on the Anninghe and Dalinagshan faults, which reaches 9–10 mm/a (Xu et al., 2008).

3.2.3 Zemuhe and Xiaojiang Fault

The Zemuhe fault extends between Xichang and Qiaojia Counties for about 110 km in length and strikes to N25° W. It is dominated by left lateral strike-slip with normal component, demonstrated by relative down throw of the northeastern block of the fault between Xichang and Puge Counties. Near Puge County, the central section of the fault has offset several branches of Heishui River, as they pass through the fault traces. The left lateral offset reaches 700–1,000 m, hence the offset geomorphic features, the beheaded abandoned gullies, were formed. The estimated late Quaternary strike-slip rate of the Zemuhe fault based on geomorphological mapping and dating of deposits is 5–9 mm/a. The recent survey shows the left-lateral slip rate of the Zemuhe fault reaches 6.4 ± 0.6 mm/a (Xu et al., 2008b).

The Xiaojiang fault is located in the Yangtze Platform that consists of early to middle Paleozoic crystalline rocks and Sinian to Triassic sedimentary rocks. The Xiaojiang fault extends 450 km southward from the Qiaojia Basin to the Red River fault (Honghe fault). The fault can be divided into the following three segments based on fault-trace geometry: The northern segment is a single strand, while the middle segment consists of two north-south striking fault strands about 15–18 km apart. In the southern segment the two fault-strands splay into a series of small faults striking north or northeast. In the middle segment, the floor of some basins is occupied by Quaternary sediments, and these include the Cangxi, Gongshan, Songming, Xundian and Yiliang Basins. Most of these basins date to the Pliocene and early Pleistocene (Yunnan Seismological Bureau, 1993; He and Oguchi, 2008). The

Xiaojiang fault runs through catchments of the Jinshajiang, Xiaojiang, Niulanjiang and Nanpanjiang rivers. These rivers and their tributaries have formed four major Quaternary fluvial surfaces (Yunnan Seismological Bureau, 1993; He and Oguchi, 2008). Some recent studies (e.g., Song, 1998) indicate that the younger fluvial surfaces of the late Pleistocene and Holocene are widely distributed along the Xiaojiang fault. The estimated late Quaternary strike-slip rate of the Xiaojiang fault based on geomorphic mapping and dating of deposits is 10 ± 2 mm/a (Xu et al., 2003).

3.2.4 Longmenshan Thrust Belt

The Longmenshan thrust belt zone is the boundary between the Qinghai-Tibetan Plateau and the rigid South China Block (Fig. 3.2), which defines the sharp and steep relief of the eastern margin of the Qinghai-Tibetan plateau. The deepest part is in the south, where mean elevation increases from ~500 m in the Sichuan basin to ~3,000 m over 50 km distance plateau-ward, and then to ~3,500 m over another 30–50 km farther northwest. The overall 400 km long Longmenshan thrust belt is NE-trending reverse faulting with right-lateral motion and consists of three major faults: the Wenchuan-Maowen fault, the Yingxiu-Beichuan fault and the Guanxian-Anxian fault. The slip rate is less than 3 mm/a, one order of magnitude lower than that on Xianshuihe fault and other major faults as detailed above. This is consistent with the fact that the Longmenshan thrust belt has been relatively quiescent in seismicity while eastern Tibet has been abundant in large earthquakes as shown in Figs. 2.1 and 3.2. The Wenchuan earthquake occurred as a surprise and needs further investigation.

3.3 Computational Method

Earthquakes have been recognized as resulting from stick-slip frictional instabilities along the faults between deformable rocks. To investigate the occurrence and rupture of earthquakes, an in-house finite element code has been developed with an arbitrarily shaped contact element strategy, named the node-to-point contact element strategy proposed with a static-explicit algorithm. It was previously applied to model of the friction contact behaviors between deformable rocks with stick and finite nonlinear frictional slip (Xing and Makinouchi, 2002a, b) and benchmarked with the so-called sandwich fault model (Xing and Makinouchi, 2002c), single fault bend model (including both the interplate and intraplate cases) (Xing et al., 2004), multiple fault bends model and further applied in simulating the South Australian interacting fault model system and South California fault system (Xing et al., 2006, Xing and Mora 2006, 2007a, 2007b; Xing and Zhang 2009). It has been applied here to investigate the Chuan-Dian fault system for an improved understanding of Wenchuan earthquake.

3.4 Numerical Analysis of Earthquake Activities

3.4.1 Computational Model

As described above, the Chuan-Dian fault system is one of the mostly active seismic regions and thus various researches on figuring out the faulting have been carried out. With detailed fault data provided by the active fault research team in Institute of Geology based on field observations, advanced digital images as well as geological knowledge of the region, we constructed a 3-D fault geometry model within a block with dimensions of about $1130 \times 750 \times 25$ km^3 (Fig. 3.3). This involved editing and smoothing the related curves/lines/surfaces defining the faults.

In order to easily generate the unstructured mesh and specify the conditions necessary for the finite-element simulation (i.e., boundary conditions and information about faults), the entire geometric model of faults (Fig. 3.3) was firstly divided into several different geometrical components representing components of the solid model and these were then used to generate finite-element meshes (e.g., see Fig. 3.4). Finally, the finite-element meshes generated for the different components were assembled together with "welding" (node equivalent), or our stick contact algorithm after meshing (Fig. 3.5). As shown in Figs. 3.4 and 3.5, more regular and fine meshes are used around the faults, while a coarse mesh is used in the other regions. This approach enables more accurate computational results to be obtained with the available and finite computational resources. The discretised model after optimization currently includes 89,622 nodes and 71,685 8-node hexahedron elements with ten faults in total (Figs. 3.3, 3.4 and 3.5).

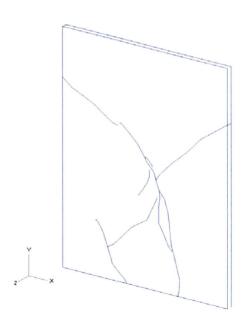

Fig. 3.3 The major faults in Chuan-Dian fault system to be analysed (refer to Fig. 3.2 for fault names)

3.4 Numerical Analysis of Earthquake Activities

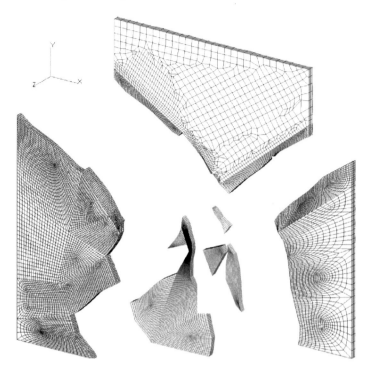

Fig. 3.4 Generated finite-element meshes after the whole fault system was divided into different geometrical components with the constraints of fault geometries

3.4.1.1 Fault System Model

Ten faults are involved here: Ganzi-Yushu fault, Xianshuihe fault, Daliangshan fault, Anninghe fault, Zemuhe fault, Xiaojiang fault, Red River fault, Lijiang-Xiaojinhe fault, Yulongxi fault and Longmenshan thrust belt. The former six faults behave as a huge left-lateral strike-slip active fault system form the eastern boundary of Chuan-Dian rhombic block in south-western China and thus denoted as the eastern boundary of Chuan-Dian rhombic block. Stick-slip algorithm with rate- and state-dependent friction law is applied in all the above ten faults to describe the fault behaviors (Xing and Mora 2006; Xing et al., 2006, 2007a and 2007b). The widely applied rate- and state-dependent friction law, which was proposed by Dieterich (1978, 1979) and Ruina (1983), is used to describe the complex phenomena along the above faults:

$$\mu = \tau/f_n = \mu_0 + \varphi + a\ln\left(V/V_{ref}\right), \quad d\varphi/dt = -\left(V/L\left(\varphi + b\ln\left(V/V_{ref}\right)\right)\right)$$

where L is the critical slip distance; a and b are empirically determined parameters; a represents the instantaneous rate sensitivity, while a-b characterizes the long-term rate sensitivity. V_{ref} and V are an arbitrary reference velocity and a sliding velocity,

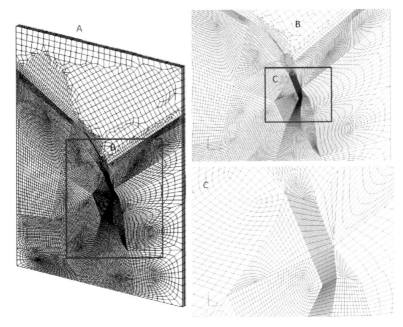

Fig. 3.5 Generated mesh for finite element analysis (**a**) total mesh after assembling with ten faults; (**b**) magnification of (**b**) and (**c**) magnification of (**c**) in (**b**)

respectively; ϕ is the state variable; f_n is the effective normal contact stress; μ_0 is the friction coefficient at reference velocity V_{ref}. For a three-dimensional case, the above-mentioned special friction law expressed can be used by simply replacing V with the relative velocity $\dot{\bar{u}}_{eq}^{sl}(=\sqrt{\dot{\bar{u}}_m \dot{\bar{u}}_m})$. The friction law is utilized here with the following parameters: $\mu_0 = 0.50$, a = 0.050, b = 0.075, $d\varphi/dt = 0$. From the above, only the friction law used here is time-dependent, thus we choose V_{ref} as the reference velocity for the whole process here.

3.4.1.2 Boundary Conditions

With reference to geological setting and GPS observation data around the Chuan-Dian fault system, the displacement constraints applied in both the external boundaries and the inside of the above computational model as detailed in Fig. 3.6. Because the motion of the Bayan Har block was resisted strongly by the South China block in the vicinity of the Sichuan basin, the fixed displacement condition is set at several internal nodes around the boundary of Sichuan basin. The detailed locations of such fixed nodes are finally defined through the relevant sensitive analysis and qualitative comparison with GPS data (Fig. 3.6).

3.4 Numerical Analysis of Earthquake Activities

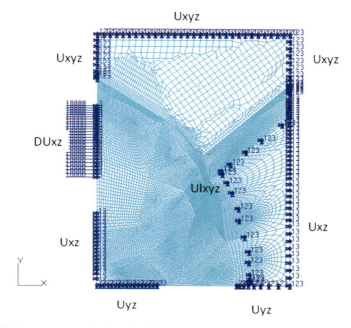

Fig. 3.6 Displacement boundaries at the *blue* marked nodes applied in the analysis. Here Uxyz and UIxyz – fixed at all the directions; Uxz – fixed along the x and z directions; Uyz – fixed along the y and z directions; DUxz – displacement applied in the positive x direction at the constant velocity of $0.01V_{ref}$ but fixed in the z direction

3.4.1.3 Material Property

All the materials have the same properties: density $\rho = 2.60$ g/cm^3, Young's modulus E = 44.8 GPa and Poisson's ratio $\nu = 0.22$.

3.4.2 Simulation Results

Assume that the initial stress is zero and the frictional parameters of all the faults are same as detailed above. With the continuous loading at the left boundary with constant velocity of $0.01V_{ref}$ along the positive direction of the x-axis, the above model of Chuan-Dian fault system is simulated and further investigated with the above finite element code. Here we focus on fault system dynamic behavior and earthquake risk assessment at the southwestern part of Longmenshan thrust belt.

To clearly illustrate the dynamic process of Chuan-Dian fault system, only snapshots of simulation results with the Y component of its velocity are shown here. The simulation results show that the fault system dynamics can be grouped into the following five stages:

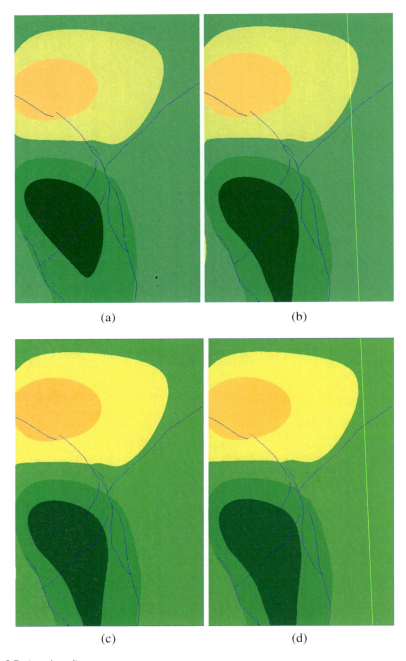

Fig. 3.7 (continued)

3.4 Numerical Analysis of Earthquake Activities 31

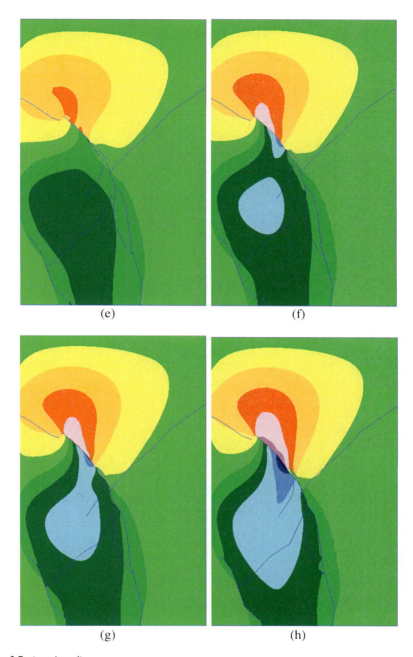

Fig. 3.7 (continued)

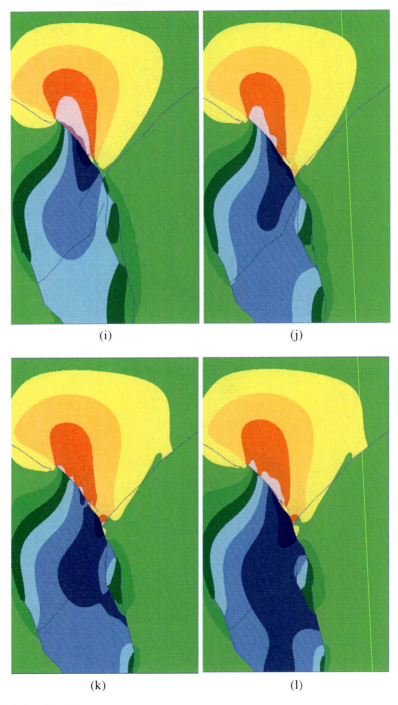

Fig. 3.7 (continued)

3.4 Numerical Analysis of Earthquake Activities

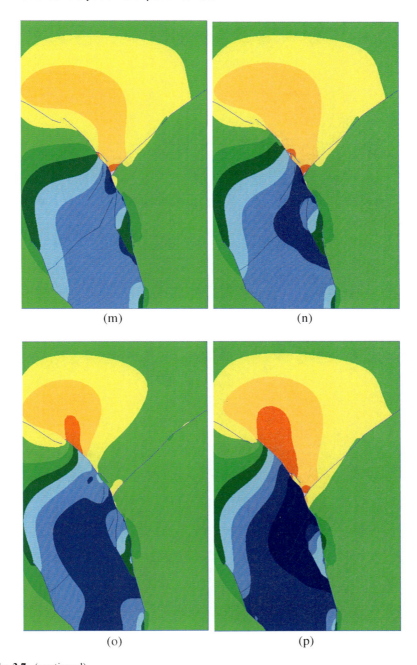

Fig. 3.7 (continued)

34　3　Earthquake Nucleation and Occurrence – Numerical Investigation

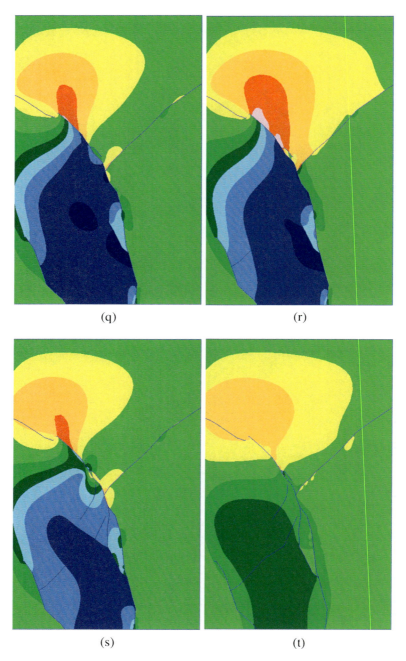

(q)　　　(r)

(s)　　　(t)

Fig. 3.7 (continued)

3.4 Numerical Analysis of Earthquake Activities

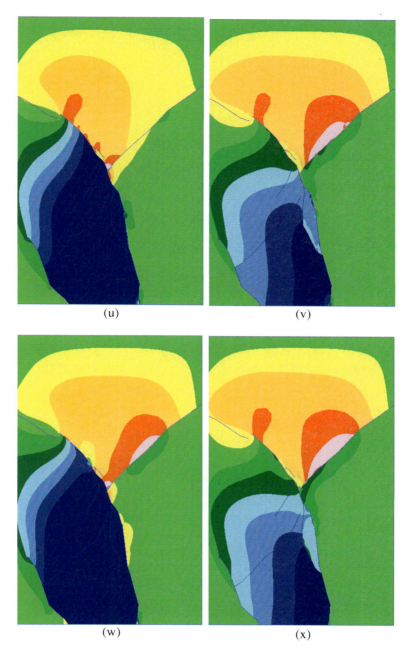

Fig. 3.7 (continued)

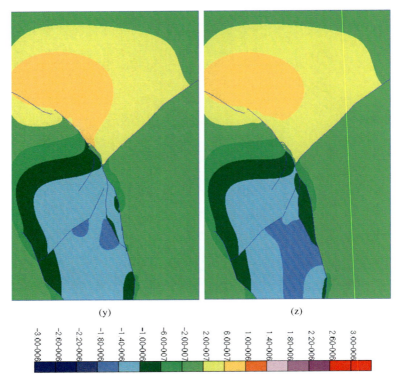

Fig. 3.7 Finite element simulation results of velocity along the y direction at the different loading stages. Loading time = $100R/V_{ref}$: (**a**) R = 0.005000; (**b**) R = 0.006730; (**c**) R = 0.010198; (**d**) R = 0.037515; (**e**) R = 0.062835; (**f**) R = 0.076834; (**g**) R = 0.10094; (**h**) R = 0.13522; (**i**) R = 0.14855; (**j**) R = 0.20060; (**k**) R = 0.23343; (**l**) R = 0.24560; (**m**) R = 0.26060; (**n**) R = 0.26756; (**o**) R = 0.29481; (**p**) R = 0.29841; (**q**) R = 0.31432; (**r**) R = 0.32121; (**s**) R = 0.32969; (**t**) R = 0.33995; (**u**) R = 0.42035; (**v**) R = 0.48214; (**w**) R = 0.51698; (**x**) R = 0.54317; (**y**) R = 5.8611; (**z**) R = 7.1111

(A) Initial deformation without rupture/slip along the faults (Figs. 3.7(a)–(d)). With the continuous loading, the fault system is deformed but no rupture/slip along the faults observed (Figs. 3.7a–d). The deformation described by the hot colors in the Bayan Har block remains exactly same during this stage, while that of the cool colors keeps the similar pattern but with gradually expending larger deformation domain in the Chuan-Dian rhombic block. That means the current loading has speeded up the motion downwards for the region in the left side of the eastern boundary (i.e. the Chuan-Dian rhombic block), but no obvious effects on the others (i.e. the right side of eastern boundary including the Longmenshan thrust belt region). It is dominated by continuous deformation with no rupture along the faults at this stage.

(B) Rupture initiation and propagation along the Xianshuihe fault (Figs. 3.7e–h). With the continuous loading and stress accumulation as above, rupture is firstly initiated at the northwestern part of Xianshuihe fault (Fig. 3.7e) and

3.5 Conclusions and Discussions 37

further propagated towards southeastern (Figs. 3.7f–h). This speeds up the motion/deformation at both sides of the eastern boundary (i.e. Bayan Har and Chuan-Dian rhombic blocks). The more loading, the larger velocity/more deformation in both blocks including those around the Longmenshan region, but no rupture along the Longmenshan thrust belt is initiated yet at this stage.

(C) Block separation and motion with the constraints of the ruptured eastern boundary (Figs. 3.7i–m). With the continuous loading and ruptured Xianshuihe fault, rupture is further developed along the other faults (e.g. Fig. 3.7i). The ruptured eastern boundary separates the Bayan Har block from the other two blocks (i.e. Chuan-Dian rhombic block and Sichuan Basin of South China block). The stress/energy release due to the ruptured faults especially at Xianshuihe results in localized motion/deformation around the relevant faults (Figs. 3.7i–l) and gradually leads to the more homogenous deformation distribution and healing up (motion state changed from the slip to stick) at the northwestern part of the ruptured Xianshuihe fault (Fig. 3.7m). The Longmenshan thrust belt ruptures at the both ends but no rupture occurs elsewhere. That means the stress accumulation remains continuously in the major part of Longmenshan thrust belt (i.e. around Yingxiu-Beichuan fault) despite of rupture occurrence at its both ends at this stage.

(D) Rupture/slip recurrence along Xianshuihe fault (Figs. 3.7n–u). With the continuous loading and partly healing up of the ruptured Xianshuihe fault at its northeastern part (Figs. 3.7m, n), the healed part of Xianshuihe fault (Fig. 3.7n) ruptures again. This leads to healing up and recurrence of rupture at both ends of Longmenshan thrust belt (Figs. 3.7n–p). Such rupture recurrences here or there along Xianshuihe fault and its effects (stress accumulation) on Longmenshan thrust belt continue cyclically (Figs. 3.7q–u) and finally lead to rupture along the major Longmenshan thrust belt (Fig. 3.7v).

(E) Rupture/slip along the Longmenshan fault (corresponding to Wenchuan earthquake) and afterwards (Figs. 3.7v–z). With the above continuous loading and the effects of rupture recurrence along Xianshuihe fault on stress accumulation around Longmenshan thrust belt, the major part of Longmenshan thrust belt finally gets ruptured (Figs. 3.7v–y), which corresponds to the occurrence of Wenchuan earthquake. With the further calculation/loading, the Xianshuihe fault gets back to the stick state again (Fig. 3.7y) and the deformation in the whole system gets more homogenous than that in the previous stages except the stage (1) (Figs. 3.7y, z).

3.5 Conclusions and Discussions

The Chuan-Dian fault system includes the following three major blocks: Bayan Har block, Sichuan basin of South China block and Chuan-Dian rhombic block. From the above simulation results, we can see:

(A) The rigid Sichuan basin strongly resists the motion/deformation of Chuan-Dian fault system during the whole process, which forces the fault system deformed roughly in Bayan Har and Chuan-Dian rhombic blocks with the opposite direction along the y direction, i.e. the Bayan Har block moves upwards while the Chuan-Dian rhombic block moves downwards subjected to the prescribed loading conditions (Fig. 3.6);

(B) The Xianshuihe fault gets ruptured which leads to localized deformation around the ruptured region and further separates the Chuan-Dian rhombic block from the Bayan Har block. Moreover, the further ruptured eastern boundary separates Chuan-Dian rhombic block with the rigid Sichuan basin of South China block. That leads to the obvious separated block motion phenomena except the major Longmenshan region and a serial of recurrent ruptures along the Xianshuihe fault. This further speeds up the stress accumulation along the Longmenshan thrust belt and finally get Longmenshan thrust belt ruptured (i.e. the Wenchuan earthquake). That is in consistent with the recorded earthquake dynamics in this region so far (Figs. 2.1 and 3.2).

(C) It is unlikely a major strike earthquake would occur in the currently silent southernmost part of Longmenshan thrust belt in the near future, because it already ruptured and no much stress accumulated along such a ruptured zone before the major rupture of Longmenshan thrust belt (i.e. the Wenchuan event). This may be in consistent with that a few of M6.0 earthquakes occurred in this area before the Wenchuan earthquake. It is different from other researchers (e.g. Toda et al., 2008; Parsons et al., 2008) and thus might need further investigation in more details.

Chapter 4
Earthquake Surface Ruptures

4.1 Introduction

The Wenchuan earthquake produced surface ruptures with the most complicated structures and the greatest length as compared with the intra-plate reverse faulting earthquakes that have ever been reported so far (Xu et al., 2008a). The earthquake ruptured simultaneously both the NE-trending, NW-dipping Beichuan and Pengguan faults with right-slip component on the middle segment of the Longmenshan thrust belt, resulting in two sub-parallel imbricate surface rupture zones accompanied by a secondary NW-trending rupture zone linking the two major rupture zones at the southern end of the Pengguan fault through a lateral ramp (Fig. 4.1). Among them, the one that is about 240 km long and extending along the Beichuan fault is named Beichuan surface rupture zone. It is the main surface rupture zone of the Wenchuan earthquake, dominated mainly by right-slip reverse faulting, and can further be divided into the Hongkou-Qingping segment dominated by thrusting with right-slip component and the Beichuan-Nanba segment dominated by right-lateral strike slip with vertical thrusting component. Among them, the Hongkou-Qingping segment has a maximum vertical displacement of 6.2±0.5 m and average vertical displacement of 3∼4 m, while the Beichuan-Nanba segment has a maximum vertical displacement of 6.5±0.5 m and a maximum right-slip displacement of 4.9±0.2 m with an average value of 2∼3 m (Fig. 4.2). The surface rupture zone that is about 72 km long and distributed along the Pengguan fault is called Hanwang surface rupture zone, dominated mainly by pure reverse faulting with a maximum vertical displacement of 3.5±0.2 m. The NW-trending surface rupture zone of about 7 km length in between the afore-mentioned two zones in the west is called the Xiaoyudong surface rupture zone. It is a reverse strike-slip fault connecting the western segment of the two imbricate main rupture zones, having a maximum left-slip and vertical displacement of 3.5 m. No coseismic surface rupture was found along the Wenchuan-Maowen fault.

The geometric structures of the Wenchuan earthquake surface ruptures are rather complicated, consisting of a mix of various rupture units including simple thrust scarp, hanging-wall collapse scarp, simple pressure ridge, dextral pressure ridge, fault-related fold scarp, back-thrust pressure ridge, pavement suprathrust scarp and local normal fault scarp (Figs. 4.3, 4.4 and 4.5). The simple thrust scarp (Fig. 4.3a)

H. Xing, X. Xu, *M8.0 Wenchuan Earthquake*, Lecture Notes in Earth Sciences 123, DOI 10.1007/978-3-642-01901-2_4, © Springer-Verlag Berlin Heidelberg 2011

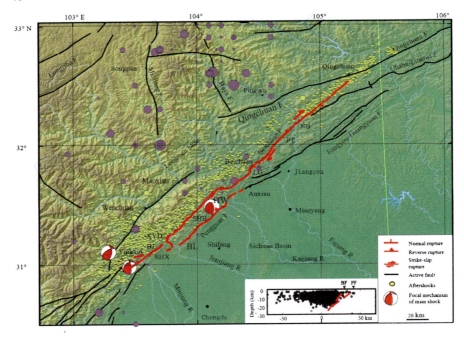

Fig. 4.1 Surface ruptures associated with the 2008 Wenchuan earthquake along the Beichuan and Pengguan faults. Historical destructive earthquakes are shown in *pink circles*, and aftershocks of the 2008 earthquake in *yellow circles*. BF: Beichuan fault, BL: Bailu Town, BJ: Bajiao Temple, HW: Hanwang Town, LG: Leigu Town, PT: Pingtong Town, NB: Nanba Town, SHB: Shaba Village, SHX: Shenxi Village, XYD: Xiaoyudong Town. (after Xu et al., 2009a)

occurred only at Bajiao Temple (N31.14522°N, E103.69189°) where a 4-m-high thrust scarp, in the mudstone, strikes N40ºE and dips 76° towards the northwest (Fig. 4.4a). On the hanging wall of the newly exposed fault plane, two sets of slickensides plunging to the west can be identified. The steeply-dipping one has a plunge of 75° at the lower part, and turns to 80° upward, to crosscut the gentle-dipping one, which has a plunge of 32°~46° (Fig. 4.4b). Generally, the over-hanging part of the newly formed thrust scarp quickly collapses onto the footwall to form a hanging-wall collapse scarp (Figs. 4.3b and 4.4c). Tensile cracks are common in the hanging-wall (Fig. 4.4d).

Simple and dextral pressure ridges are widely developed (Figs. 4.3c and d). Where such deformation occurs, it is common that near-surface material such as turf, soil and concrete pavement was dragged along and that it overrides the ground surface on the hanging-wall block (Figs. 4.4e and f). Vegetation on the top of the scarp, such as crops and trees, tends to follow the scarp geometry and to be tilted accordingly, to the limit where trees would fall. This forms the so called "drunkard woods" (Fig. 4.4e).

A fault-related fold scarp describes a scarp formed by folding of near surface strata in response to a blind reverse fault (Fig. 4.3e). For example, the sub-horizontal yellowish brown terrace deposits at the northern bank of the Jianjiang River, in

4.1 Introduction

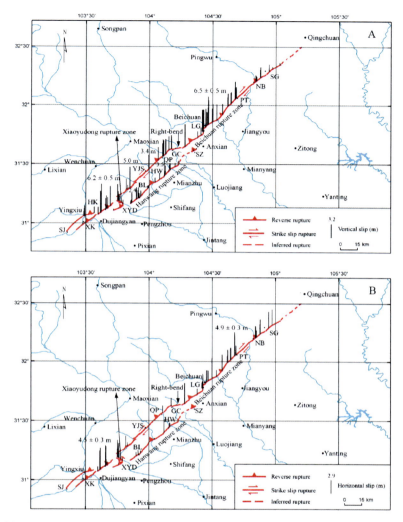

Fig. 4.2 (**A**) Along-strike distribution of measured coseismic vertical offsets, and (**B**) Horizontal offsets along the Beichuan, Pengguan and Xiaoyudong rupture zones. BL: Bailu Town, GC: Gaochuan Town, QP: Qingping Town, HW: Hanwang Town; HK: Hongkou Town, LG: Leigu Town, NB: Nanba Town, PT: Pingtong Town, SG: Shuiguan Town, SZ: Sangzao Town, XK: Xuankou Town, XYD: Xiaoyudong Town (After Xu et al., 2009a)

Beichuan Town, were folded toward the SE, resulting in a 3.1-m-high scarp at the ground surface, with no apparent fault. The cement pavement was broken during the earthquake (Fig. 4.4g), and the yellow-colored centerline of the road was offset by 2.4 m right-laterally across the fold scarp.

The back thrust scarp is named for a secondary reverse fault scarp developed on the hanging wall of the main reverse fault scarp, with an opposite dip direction

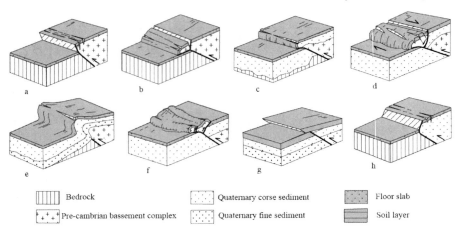

Fig. 4.3 Fault-scarp features along the Beichuan and Pengguan faults of the middle Longmenshan thrust belt, eastern Tibetan. (**a**) simple thrust scarp; (**b**) hanging-wall collapse scarp; (**c**) Simple pressure ridge; (**d**) dextral pressure ridge; (**e**) fault-related fold scarp; (**f**) back-thrust pressure ridge; (**g**) pavement suprathrust scarp; (**h**) local normal fault scarp (After Yu et al., 2009)

(Fig. 4.3f). Linear warping occurs in between the main and secondary scarps, and secondary tension cracks sub-parallel to the scarp could develop on the warp. The pavement suprathrust scarp occurs often where concrete pavement crosses the scarp. It is due to underground thrusting and shortening under a more rigid superficial layer (Figs. 4.3g and 4.4h).

Additionally, local normal fault scarps are also observed for ~7 km north of Beichuan Town and for ~500 m at Sujiayuan Village, Shikan Town, Pingwu

Fig. 4.4 Photos showing Fault-scarp features of the Wenchuan earthquake. (**a**) a 4-m-high simple thrust scarp at Baojiao Temple (N31.14522°N, E103.69189°), where fault strikes N32°E and dips 76° NW, with slickenside striations (**b**), showing this fault is dominated by reverse faulting along the Hongkou sub-segment of the Beichuan rupture zone. (**c**) a 4-m-high hanging-wall collapse scarp at Baojiao Temple (N31.14522°N, E103.69190°) less than 10 m to the east of the simple thrust scarp. (**d**) tension cracks on a 2.7-m-high hanging-wall collapse scarp at Shenxigou Village (N31.08925°, E103.61492°) along the Hongkou sub-segment of the Beichuan rupture zone. (**e**) simple pressure ridge ~3 m high pressure ridge and drunkard woods at Xiaoyudong Town (31.19575°N, 103.75269°E) along the Xiaoyudong rupture zone. (**f**) dextral pressure ridge ~1.5 m high with a maximum right-lateral slip of ~4.9 m on the youngest terrace of Pingtong River at a site (N 32.05403°, E 104.67725°), south Pingtong Town, along the Nanba sub-segment of the Beichuan rupture zone. (**g**) fault-related fold scarp about 3.1 m high at Beichuan Town (31.82894°N, 104.45689°E) along the Nanba sub-segment. (**h**) pavement suprathrust scarp with a 1.4-m overlap at Longquan Village (32.28697°N, 104.94619°E) along the Hanwang rupture zone. (**i**) a 1.7-m-high normal fault scarp that strikes N52°E and dips 46°SE at Sujiayuan Village (N31.83269°, E104.94619°), Shikan Town, Pingwu County. (**j**) slickenside striations on the normal fault plane rake 22°~46° west, showing that the fault is dominated by right-lateral faulting with a normal component (After Yu et al., 2009)

4.1 Introduction

Fig. 4.4 (continued)

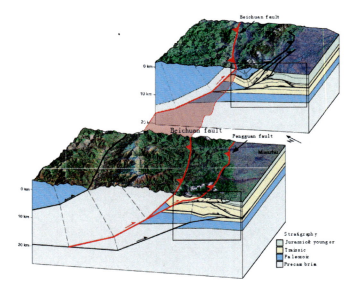

Fig. 4.5 The Wenchuan earthquake algorithm and the ruptured faults (in *red*)

County (Figs. 4.3h, 4.4i and j). At both sites, the northwestern side consists of Cambrian mudstone and the southeastern side is formed of Carboniferous limestone and alluvial-fluvial slumping deposits. During the Wenchuan earthquake, crustal shortening on the reverse fault forced the softer Cambrian mudstone and alluvial-fluvial slumping deposits up onto the stronger Carboniferous limestone. Back-tilting occurred at the fault tip, near the ground surface, and a local normal fault scarp has formed.

Moreover, numerous tension or tension-shear cracks striking along the strike of the surface ruptures occur on the top of the simple thrust scarp, hanging-wall collapse scarp, simple pressure ridge, dextral pressure ridge, fault-related fold scarp and local normal fault scarp. Most of them extend vertically downward, and can be assigned to secondary tension fractures produced under local tensional stress condition on the warp of the hanging wall of the reverse fault. Field observations discovered also that a series of N70°–90°E-trending tension cracks at large angle oblique to the scarp are developed on the hanging wall of the thrust scarp, indicating that the surface ruptures have a right-slip component.

The surface ruptures associated with the Wenchuan earthquake have distinct features in their rupture patterns, rupture width, geometric structures and distribution of coseismic displacement. The rupture structures are the most complicated and the rupture length is the longest for reverse faulting with right-slip component events that ever reported. The field investigations were focused on the surface ruptures and described respectively in the following three ruptures zone: Beichuan, Hanwang and Xiaoyudong. The detailed measurement data are listed in the Appendices 4.1–4.3.

4.2 Beichuan Surface Rupture Zone

The Beichuan rupture zone developed along the Beichuan fault is the main one among the three surface rupture zones produced by the Wenchuan earthquake. The surface rupture initiates from the west of the Yingxiu Town, Wenchuan County (31.061°N, 103.333°E) in the west close to the epicenter or the initial rupturing point located by the China Seismic Network Center and the U. S. Geological Survey, and terminates in the east of Shuiguan village, at the border between the Pingwu and Qingchuan Counties (32.233°N, 104.878°E) in the east, having a total length of 240 km (Figs. 4.5–4.112). On the western end of the rupture in the vicinity of Yingxiu Town, there exist two N60°–70°E-trending secondary rupture zones (Fig. 4.1), appearing as simple thrust rupture or deformation on the surface. Among them, the northern branch obliquely cuts across north Yingxiu Town along NEE-SWW direction, resulting in a 1–2.3 m high pressure ridge. No significant strike-slip displacement of linear makers, such as highway, river channel and terrace across the scarp, was observed, indicating that the Wenchuan earthquake initiated from compressive thrust type rupturing at its source (Figs. 4.6, 4.7 and 4.8). At the site several kilometers to the east of Yingxiu Town, the two secondary rupture zones has merged into a N42°±5°E-striking single strand with right-slip component, which is becoming greater eastward (Figs. 4.9–4.52). In the north of Qushan Town, Beichuan County, the right-slip component is significantly greater than the vertical slip component (Fig. 4.2), in accordance with the kinetic features as indicated by the plunge of fault strike on slickenside observed at Bajiaomiao, Hongkou Village of Dujiangyan City and in the north Qushan Town of Beichuan County on the western segment of the zone. At Bajiaomiao observation point, the newest fault plane strikes N40°E and dips northwest at an angle of 76° (Figs. 4.18, 4.19 and 4.20). On the fault plane of the hanging wall, two sets of slickenside striation plunging to the west were observed. Among them, the steeply-plunging one with a rake of 75° at the lower part and 80° upward crosscuts the gently-dipping one, which has a rake of 32°~46° (Fig. 4.19). This may indicate that the earthquake rupturing was dominated mainly by thrusting with a small amount of right-slip, resulting in eventually a 4 m-high reverse fault scarp. At the observation site to the north of Qushan Town, the newest fault plane strikes N55°E and dips southeast at an angle of 52°~90° (Figs. 4.53–4.67). On the fault plane of the foot wall, sub-horizontal slickenside striation with a rake of about 25° is well developed (Fig. 4.56). This may indicate that the site is dominated mainly by right-lateral strike-slip with normal dip-slip component as indicated by the rising of the northwest wall and the descending of the southeast wall. Since the site as a whole is dominated by the rising of the northwest wall, it is postulated that the SE-dipping normal dip-slip surface rupture, extending for several kilometers in the north of Qushan Town, might be converted into a NW-dipping reverse fault at depth.

Geometrically, the Beichuan rupture zone consists of a mix of en echelon arranged rupture units, including the Hongkou, Longmen Shan-Qingping and Beichuan-Nanba secondary rupture zones. Among them, the Hongkou secondary rupture zone is about 39 km in length, and is a segment dominated mainly by

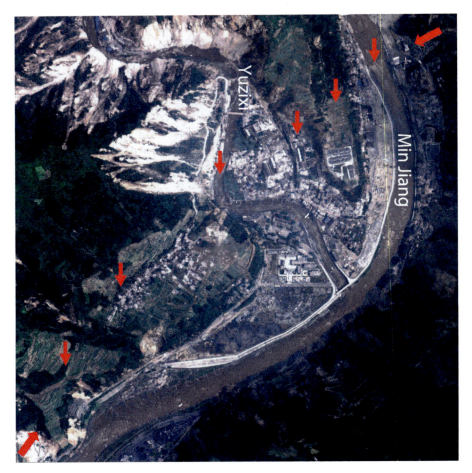

Fig. 4.6 Air photograph of Yingxiu Town, Wenchuan County after the Wenchuan earthquake (the *red solid arrow* indicates the position of surface rupture zone, and the position of the surface rupture zone at the Minjiang River valley, Dujiangyan-Wenchuan highway, and Yuzixi observation points)

thrusting. The maximum vertical displacement associated with the Wenchuan earthquake has been measured to be 6.2±0.5 m by using total station and 3D scanner at site to the north of Shenshuiwan power station at Shenxigou village across a N50°E-trending pressure ridge (Fig. 4.17), while the vertical offset in the adjacent areas is in the range of 4~5 m (Figs. 4.10, 4.11, 4.12, 4.13, 4.14, 4.15, 4.16, 4.17 and 4.18). In addition, right-slip component was observed also at individual sites. For example, right-slip of 4.5 m and vertical slip of 2.7 m of a cement pavement were measured at Shenxigou (Fig. 4.13). Taking the dip angle of the reverse fault as 47°, which is deduced from the envelope of the concentration belt of the relocated aftershocks, we may estimate the maximum crustal shortening across the Hongkou segment to be about 5.8 m.

4.2 Beichuan Surface Rupture Zone

Fig. 4.7 Simple pressure ridge formed along the Yuzixi River and the road on its western bank to the northwest of Yingxiu Town, Wenchuan County. The strong deformation of the hanging wall caused the collapse of the slope and roadbed. Viewed to northwest

Fig. 4.8 The Beichuan fault co-seismically offset the Minjiang River valley and its multiple levels of terraces, resulting in simple pressure ridge and waterfall scarp. Near surface tension cracks occurred on the hanging wall of the scarp, and the residential buildings on the surface rupture zone were completely destroyed; a 6-storey building under construction on the foot-wall of the zone near the fault was damaged in its structures and the second floor was missed. Viewed to west

Fig. 4.9 The Minjiang River valley and the road from Dujiangyan to Wenchuan on the T1 terrace on the western bank of the Minjiang River to the north of Yingxiu Town, Wenchuan County were vertically offset to form 2.3 m-high simple pressure ridge. Brittle deformation occurred on the cement pavement of the road. Viewed to north

Fig. 4.10 The hanging wall of a fault-related fold scarp and cement pavement of a road were SE-tilted, resulting in typical "drunkard woods" at Shenxigou, Hongkou Town, Dujiangyan City

4.2 Beichuan Surface Rupture Zone

Fig. 4.11 The fault-related fold scarp has caused the SE-tilting of the cement pavement of a road at Shenxigou, Hongkou Town, Dujiangyan City. The dip angle may reach up to 35~40°, and the car on the road inclined with the inclination of the ground. Viewed to southwest

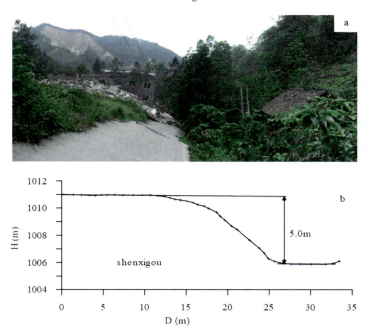

Fig. 4.12 An about 5 m-high fault-related fold scarp was developed on the cement pavement of a road at Shenxigou, Hongkou Town, Dujiangyan City. The houses built across the scarp were collapsed; the houses on the hanging wall on the left side were damaged in their structures, while the wood structured house on the foot-wall on the right was damaged but not collapsed. Viewed to northeast

50   4 Earthquake Surface Ruptures

Fig. 4.13 Fault-crossing cement road was offset by the hanging-wall collapse scarp at Shenxigou, Hongkou Town, Dujiangyan City, displaying a right-slip offset of 4.5 m and vertical offset of 2.7 m. Viewed to northwest

Fig. 4.14 Tension cracks perpendicular to the deformed ground surface occurred on the hanging-wall collapse scarp at Shenxigou village, Hongkou Town, Dujiangyan City. Viewed to southwest

4.2 Beichuan Surface Rupture Zone

Fig. 4.15 A hanging-wall collapse scarp developed in a corn field at Shenxigou village, Hongkou Town, Dujiangyan City. The scarp is about 3.7 m high; the corn stems on the scarp in front of the photo incline; the plants are vertical on both walls of the scarp, and the woods in distance are tilted to form so-called "drunkard woods". Viewed to northeast

Fig. 4.16 The western embankment of the Baishahe River at Zhoujiaping village, Dujiangyan City, was offset by a dextral pressure ridge. The uplift is 3.1 m and the right-slip displacement is 1.9 m. Viewed to north

52   4 Earthquake Surface Ruptures

Fig. 4.17 Simple pressure ridge at Shenxigou village, Hongkou Town, Dujiangyan City. The scarp is 6.2±0.5 m high, where the maximum coseismic vertical offset of the Hongkou-Qingping segment (western segment) of the Beichuan-Yingxiu rupture zone was obtained. Viewed to northeast

Fig. 4.18 Simple thrust scarp at Bajiaomiao village, Dujiangyan City. Viewed to northeast

4.2 Beichuan Surface Rupture Zone

Fig. 4.19 Two-period slickenside striations observed at Bajiaomiao, Dujiangyan City. The early-period striation indicates right-slip with thrust component; the latest striation indicates that the Wenchuan earthquake was dominated mainly by thrusting with right-slip component. Viewed to northwest

Fig. 4.20 Waterfall produced by simple thrust scarp at the Bajiaomiao in Dujiangyan City. The black materials are fault shattered zone. Viewed to north

Fig. 4.21 A fault-related fold scarp on a path to the Donglinsi Mine at Lonmenshan Town, Pengzhou City. The scarp is N40°E- striking and 2 m high. Viewed to north

Fig. 4.22 A warping scarp on a touring road to Xiangshuidong at Lonmenshan Town, Pengzhou City. The scarp is N20°E-striking and 1 m high with 0.8 m right-slip. Viewed to north

Fig. 4.23 A warping scarp on a pluvial platform at Xiejiadian Village, Lonmenshan Town, Pengzhou City. The scarp is N65°E-striking. Viewed to north

4.2 Beichuan Surface Rupture Zone

Fig. 4.24 The NW-wall of the foundation of the construction site of the "Chengdu's impression" buildings at Donglinsi Village, Lonmenshan Town, Pengzhou City was uplifted along N40°E direction for 2.5 m with a 2.2 m right-slip offset. Besides, collapse also occurred. Viewed to north

Fig. 4.25 Simple pressure ridge along the range front at Muguaping Village, Pengzhou City. The scarp is 3.5 m high; the vegetation on the scarp was growing up crookedly after the quake. Viewed to north-northeast

Fig. 4.26 Fault-related fold scarp of about 3.5 m height developed on the T2 terrace at the Phosphorus Mine of Qingping Village, Mianzhu City. The pine trees on the scarp are tilted to form "Drunkard woods". Viewed to northeast

Fig. 4.27 Close the fault-related fold scarp up on the T2 terrace at the Phosphorus Mine of Qingping Village, Mianzhu City. The pine trees on the scarp are tilted to form "Drunkard woods". Viewed to northeast

4.2 Beichuan Surface Rupture Zone

Fig. 4.28 *Yellowish brown* and *dark grey* fault gouges outcropped on the cross section of the earthquake surface rupture zone at the Phosphorus Mine of Qingping Village, Mianzhu City. Viewed to northeast

Fig. 4.29 Fault-related fold scarp and trench location on the western side of the highway from Mianzhu to Maoxian at Qingping Village, Mianzhu City. Viewed to northeast

Fig. 4.30 Cross section of the latest surface deformation of fault-related fold scarp developed on the terrace as revealed by the trench on the western side of the highway from Mianzhu to Maoxian at Qingping Village, Mianzhu City. The brown overburden on the bottom of the scarp is thickened, and there is another earlier fold deformation event at the lower part. Viewed to southwest

Fig. 4.31 A 3.8 m-high scarp with a right-slip offset of 1.5 m on the highway from Mianzhu to Maoxian at Qingping Village, Mianzhu City. On the background there is a swarm of landslides. Viewed to north

4.2 Beichuan Surface Rupture Zone

Fig. 4.32 The surface rupture zone passes through the col at Gaochuan Village, Mianzhu City, and the original riverbed was raised to form a waterfall scarp. Viewed to southwest

Fig. 4.33 The newest fault plane on bedrock shattered zone on the eastern side of the highway from Mianzhu to Maowen at Qingping Village, Mianzhu City. The fault plane is N10°E-dipping at an angle of 52°. Viewed to northeast

Fig. 4.34 A 3.5 m-high fault scarp on top of the newest fault plane of bedrock shattered zone, eastern side of the highway from Mianzhu to Maowen at Qingping Village, Mianzhu City. The trees on the scarp are fallen down. Viewed to southwest

4.2 Beichuan Surface Rupture Zone

Fig. 4.35 Bedrock fault and its shattered zone on the eastern side of the highway from Mianzhu to Maowen at Qingping Village, Mianzhu City. Viewed to northeast

Fig. 4.36 Simple pressure ridge about 2.2 m high and tension crack zone about 180 m wide at Quanshui Village, Gaochuan Township, Mianzhu City. Viewed to southwest

Fig. 4.37 A 3.19 m-high simple pressure ridge on the road at Quanshui Village, Gaochuan Township. Viewed to west

Fig. 4.38 Fault-related fold scarp of about 2.59 m high and "drunkard woods" on the T1 terrace at Quanshui Village, Gaochuan Township. Viewed to southwest

4.2 Beichuan Surface Rupture Zone

Fig. 4.39 The uplifted riverbed formed a 2~3 m-high coseismic scarp and waterfall at Quanshui Village, Gaochuan Township. Viewed to southwest

Fig. 4.40 Bedrock fault scarp (on the *left* of the wire pole) at Xiaojiaqiao Village, Chaping Township, Anxian County. The scarp is 5.2 m high. Viewed to northeast

64   4 Earthquake Surface Ruptures

Fig. 4.41 Bedrock fault scarp (on halfway up the hill) at Xiaojiaqiao Village, Chaping Township, Anxian County. The scarp is 5.2 m high, and a small gully was offset right-laterally for 4.3 m. Viewed to north

Fig. 4.42 A bedding fault (near the fallen trees in distance) and a large-scale bedrock collapse in the vicinity of the fault at Xiaojiaqiao Village, Chaping Township, Anxian County. Viewed to northwest

4.2 Beichuan Surface Rupture Zone

Fig. 4.43 Simple pressure ridge on the high floodplain of the southern bank of the Laochanghe River about 4.5±0.5 m high, north of the 1st Group of Shiyan Village, Leigu Town, Beichuan County. Viewed to west

Fig. 4.44 Two imbricated N75°W-striking simple pressure ridges lower in the north and higher in the south on the ground of a pigsty at a site to the east of the 1st Group of Shiyan Village, Leigu Town, Beichuan County. The total height of the pressure ridges is 3.7 m, and the right-slip offset is 2 m. Viewed to southwest

Fig. 4.45 The current riverbed and low terrace was right-laterally thrusted to form a simple pressure ridge at Wuxing Township, Mianzhu City. Viewed to southwest

Fig. 4.46 Fault-related fold scarp on the low terrace at Wuxing Township, Mianzhu City. Viewed to southwest

4.2 Beichuan Surface Rupture Zone

Fig. 4.47 A path was offset to form a simple pressure ridge at Maliuwan Village, Wuxing Township, Mianzhu City. Viewed to southeast

Fig. 4.48 The T2 terrace and the cement road on it was offset by the earthquake fault to form a 2~3 m-high simple pressure ridge in the north of the 1st Group of Shiyan Village, Leigu Town, Beichuan County. The houses on the hanging wall near the fault were completely destroyed, and the trees incline toward the foot wall. Viewed to southeast

Fig. 4.49 The T1 terrace on the south bank of the river was offset by earthquake fault to form three N25°E-trending simple pressure ridges higher in the west and lower in the east at Liulin Village, Leigu Town, Beichuan County. The vertical offset of the eastern main scarp is 2.3±0.3 m, offset left-laterally the earth bank between fields for 1.2±0.1 m. The vertical offsets of the two small scarps in the west are 0.9±0.1 m and 0.8±0.1 m, respectively. Viewed to north

Fig. 4.50 The current riverbed was offset to form two N25°E-trending scarps at Liulin Village, Leigu Town, Beichuan County. The vertical offset is 4.2±0.4 m. Viewed to southwest

4.2 Beichuan Surface Rupture Zone

Fig. 4.51 A N60°E-trending landslide back-fringe scarp about 5 m high was formed on the T1 terrace on the west side of a gully at the 2nd Group of Pinhshang Village at the head of Zhaojiagou gully (S30°E-extending), Leigu Town, Beichuan County. Viewed to north

Fig. 4.52 A pressure ridge offset the gully bed and the T1 terrace on the east side of the gully, resulting in a NNE-trending 3-m-high scarp at the 3rd Group of Pingshang Village, Mofanggou, Leigu Town, Beichuan County. The pressure ridge right-laterally offset simultaneously the water pipeline, field banks and cement road for about 2.1 m. Viewed to southwest

Fig. 4.53 The Anxian to Beichuan highway was offset by the earthquake fault to form a N20°E-trending about 3 m-high fault-related fold scarp higher in the west and lower in the east, at Renjiaping Village, 1 km to the south of the Beichuan County. Viewed to north

Fig. 4.54 The folded and broken man-made canal showing uplift of the northwest side and the warping of the cement bottom of the canal in the south of Qushan Town, Beichuan County. Viewed to north

4.2 Beichuan Surface Rupture Zone

Fig. 4.55 The flexure and fold of the upper gravel deposits and the middle red-brown brick layers of the Jianjiang terrace in the vicinity of the Beichuan Grand Hotel, Qushan Town, Beichuan County. As a result, a 3.1 m-high fault-related fold scarp was formed, causing the brittle fracturing of the cement pavement of the highway, the central line of which was offset right-laterally for 2.4 m. Viewed to NNW

Fig. 4.56 The newest local normal fault plane is outcropped on the foothills on the eastern side of the north Qushan Town, Beichuan County. The fault plane strikes N56°E, SE-dipping at angle of 52°~90°. Sub-horizontal slickenside striations are well developed on the footwall of the fault, west-plunging with a rake of about 25°. This may indicate that the observation site is dominated mainly by righ-lateral strike-slip with a normal dip-slip component as indicated by the raising of the northwest wall and the descending of the southeast wall. The measured vertical offset is 1.7 m, and the right-slip offset is estimated from the rake of the striation to be about 3.7 m. Viewed to northeast

Fig. 4.57 Earthquake fault slip resulted in a local normal fault scarp and a dammed lake on its eastern side at Maoba Village, Qushan Town, Beichuan County, "Drunkard woods" also occurred in the vicinity of the fault. Viewed to northeast

Fig. 4.58 The SE-dipping local normal fault scarp was developed on the foothills on the eastern side of the north Qushan Town, Beichuan County, resulting in trough-in-slope landform, but it still indicates the uplift of the northwest block. Viewed to northeast

4.2 Beichuan Surface Rupture Zone

Fig. 4.59 A SE-dipping secondary back-thrust fault was developed on the upthrown side of the antithetic normal fault scarp to the north of Qushan Town, Beichuan County. This may indicate that the formation of normal fault scarp was closely related to the compression from NW to Se. The secondary back-thrust fault has tensional component, indicating that late-stage collapse occurred along the slope. Viewed to southwest

Fig. 4.60 Antithetic normal fault scarp in the north of Qushan Town. Viewed to southwest

Fig. 4.61 Steeply dipping bedrock fault plane outcropped at Shaba Village to the north of Beichuan County town. Viewed to southwest

Fig. 4.62 Bedrock fault scarp at a site in between Shaba Village and north Qushan Town. The scarp is about 4 m high; on the fault plane South-plunging slickenside striation with a rake of 30°~40° was well developed, indicating that the fault was dominated by right-lateral strike-slip. Viewed to southwest

4.2 Beichuan Surface Rupture Zone

Fig. 4.63 Slickenside and striation on bedrock fault plane at Shaba Village to the north of Beichuan County. Viewed to southwest

Fig. 4.64 The maximum vertical offset at Shaba Village to the north of Qushan Town reaches up to 6.5±0.5 m. The upper part of the scarp, where bamboos and trees are standing, is the indicator of the preexisting fault scarp landform. Viewed to southwest

Fig. 4.65 The cement road was offset right-laterally 3.3 m and vertically 6 m at Shaba Village to north of Beichuan County. Viewed to northwest

Fig. 4.66 Two step-like NE-trending and SE-dipping normal fault scarps on the loess platform at Heshangping Village, Beichuan County. The total height of the scarps is 3.2 m; the scarps dissect the loess-like near-surface overburden and the underlying bedrocks; eyebrow-like ridge occurs on the right of this picture owing to the occurrence of about 4 m right-slip offset. Viewed to northwest

4.2 Beichuan Surface Rupture Zone

Fig. 4.67 A NW-trending trench excavated across the eastern step-like normal fault scarp on the loess platform at Heshangping Village. The trench reveals the *grayish yellow* fault gouge materials developed on the newest fault plane between the *dark grey* broken mudstone and *yellowish brown* stone-bearing soil bed. Viewed to southwest

Fig. 4.68 Nearly N-S-trending fault-related fold scarp developed on the high terrace on the eastern side of a road at Dengjia Village, Beichuan County. The scarp is 5.6±1.0 m high, which is the accumulated height of the original slope and two seismic deformation events. Viewed to SSE

Fig. 4.69 Fault-related fold scarp developed on terraced field at the turning point, where the fault-related fold scarp turns it strike from nearly E-W-trending to NE-trending on hillslope on the eastern side of a road at Dengjia Village, Beichuan County. The trees on the scarp were tilted to form "drunkard woods". Viewed to SSW

Fig. 4.70 Nearly N-S-trending fault-related fold scarp developed on the western side of a road at Dengjia Village, Beichuan County. The scarp is 4.1 m high; the trees on the scarp were tilted, and the house were collapsed. Viewed to south

4.2 Beichuan Surface Rupture Zone

Fig. 4.71 Earthquake fault scarp at Huangjiaba Village, Beichuan County. The scarp is N30°~40° E-trending and 3.6±0.1 m high, having a right-slip offset of 1.4 m. It perpendicularly passes through a canal, resulting in a waterfall scarp, Viewed to northwest

Fig. 4.72 An about 2.5 m-high fault-related fold scarp developed on a corn field at Huangjiaba Village, Beichuan County. Continuous flexural folding occurred from the hanging-wall (on the *left*) to the footwall, as reflected by the vertically standing trees on the left-hand side, tilting trees on the scarp and fallen down trees at the bottom of the scarp; typical phenomenon of "drunkard woods" can be seen on the footwall. Viewed to northeast

Fig. 4.73 A 3~4 m-high fault-related fold scarp developed on the T1 terrace at Chenjiaba, Beichuan County. En-echelon tension cracks are well developed on the scarp. Viewed to northeast

Fig. 4.74 A 3~4 m-high fault-related fold scarp and waterfall scarp developed on the T1 terrace at Chenjiaba, Beichuan County. Viewed to northwest

4.2 Beichuan Surface Rupture Zone

Fig. 4.75 An about 2.3 m-high fault-related fold scarp developed on farmland at Fenghuang Village, Guixi Township, Beichuan County. The scarp is superposed on the preexisting scarp. The slope angle of the scarp is 25°, and the crops and trees on it were tilted. Viewed to northwest

Fig. 4.76 The earth bank between field was right-laterally offset for 1.6 m on the T1 terrace at Fenghuang Village, Guixi Township, Beichuan County. Viewed to southeast

Fig. 4.77 A 2.2~2.5 m-high and 40°~43°-trending fault scarp is nearly perpendicular to the terrace and floodplain on the western bank of a stream at Fenghuang Village, Guixi Township, Beichuan County. Viewed to southwest

Fig. 4.78 A N40°E-trending fault-related fold scarp at the 1st Team of the Fenghuang Village, Guixi Township, Beichuan County. The scarp is sub-perpendicular to the cement road; the northwest side of the scarp was raised and the southeast side was descended; the slope is about 7 m long, and the slope angle is 19°~20°; the vertical offset is 2.4 m and the right-slip displacement is 2.8~3 m. Viewed to northwest

4.2 Beichuan Surface Rupture Zone

Fig. 4.79 A path on the T1 terrace of the Pingtong River was offset right-laterally 4.9 m and vertically 1.5 m at the southeastern side of Muerdi Village, Pingtong Town, Pingwu County. The horizontal offset value is the maximum horizontal offset measured on the surface rupture zones associated with the Wenchuan earthquake. Viewed to northwest

Fig. 4.80 Numerous tension-shear cracks slightly oblique to the strike of the scarp developed on the hanging-wall of the scarp on the T1 terrace of the Pingtonghe River at the southeastern side of Muerdi Village, Pingtong Town, Pingwu County. This may indicate that the scarp, on which the maximum right-slip offset was measured, is a right-lateral strike-slip fold scarp. Viewed to southeast

Fig. 4.81 The earth bank between fields on the T2 terrace of the Pingtong River at the southeastern side of Muerdi Village, Pingtong Town, Pingwu County was offset right-laterally 4.4~6.0 m. The vertical offset is 3.3 m (including two seismic deformation events); the excavated trench across the field bank reveals paleo-earthquake event. Viewed to northwest

Fig. 4.82 Trench excavated across the scarp on the T2 terrace of the Pingtong River at the southeastern side of Muerdi Village, Pingtong Town, Pingwu County reveals 3 paleo-earthquake events (including the Wenchuan earthquake of May 12). Viewed to southwest

4.2 Beichuan Surface Rupture Zone

Fig. 4.83 Right-lateral strike-slip fold scarp and en-echelon tension-shear cracks on its hanging-wall at the middle part of Muerdi Village, Pingtong Town, Pingwu County. 3-dimensional scanning image and measured profile across the scarp show that the scarp is 3.7 m high, and the right-slip offset is 3.7 m. Viewed to northwest

Fig. 4.84 The NE-trending fault-related fold scarp to the north of Muerdi Village, Pingtong Town, Pingwu County blocked up the channel of the nearly SD-N-running Pingtong River to form a local sag pond. The white band without grass in this figure is a path penetrating into the sag pond. Viewed to southwest

Fig. 4.85 The N50°E-trending fault-related fold scarp in the north of Pingtong Town, Pingwu County obliquely cuts the nearly S-N-extending riverbed of the Pingtong River and the river island to form river channel scarp and waterfall; the 3-dimensiomal scanning image and the measured profile across the scarp show that the scarp is about 2.5 m high. Viewed to SWW

Fig. 4.86 The N50°E-trending fault-related fold scarp in the north of Pingtong Town, Pingwu County obliquely cuts the eastern dykes of the nearly S-N-extending riverbed of the Pingtong River. The scarp is dominated by both vertical offset and right-slip. Viewed to north

4.2 Beichuan Surface Rupture Zone

Fig. 4.87 The surface of T1 terrace of the Pingtong River was warped to form a an about 1.5 m-high fault-related fold scarp at Pingtong Town, Pingwu County. Viewed to northwest

Fig. 4.88 The northwest side of a fish pond near the T1 and T2 terraces at Pingtong Town, Pingwu County was folded and uplifted, resulting in a south-sloping and about 2 m –high fault-related fold scarp; the bank of the pond was offset right-laterally about 1.9 m. Viewed to north

Fig. 4.89 The northwest wall of the road to Pingwu County town on the T2 terrace at Pingtong Town, Pingwu County was uplifted about 2.5 m relative to its southeast wall at Pingtong Town, Pingwu County; the road was offset right-laterally for 1.5 m, and the broken road pavement indicates that the width of the surface rupture zone is 41 m. Viewed south

Fig. 4.90 Crocodile-mouth-like scarp at Pingtong Town, Pingwu County; the cement pavement of the road bulged and superposed; the superposition segment is 50–60 cm long. Viewed to northeast

4.2 Beichuan Surface Rupture Zone

Fig. 4.91 The outcrop of earthquake fault at Pingtong Town, Pingwu County; the hanging wall consists of yellowish brown weathering crust of mudstone, and the footwall of pale schist; the cross section reveals 3 sets of faults and a shattered fault zone; the occurrences of the 3 faults are 50°–60°/W $\angle$ 20°, 50°–60°/W $\angle$ 25° and 50°–60°/W $\angle$ 30°–40°, respectively; the fault plane is fresh, and eaves-like structure was formed owing to right-slip. Viewed to northeast

Fig. 4.92 Slickenside striation with a rake of 30°developed on the earthquake fault plane at Pingtong Town, Pingwu County. This may indicate that the fault was dominated by right-lateral strikr-slip with thrust component. Viewed to northwest

Fig. 4.93 Earthquake surface rupture zone at a site to the northeast of Nanba Town, Pingwu County. The rupture zone is N45°–50°E-striking; the northwest side of the road was uplifted 1.52 m and right-laterally offset 2.5±0.2 m. Viewed to southeast

Fig. 4.94 Earthquake fault outcropped on the southern side of a road at a site to the northeast of Nanba Town, Pingwu County. The fault strikes N50°E and dips northwest with an angle of 50°–60°. Viewed to northeast

4.2 Beichuan Surface Rupture Zone

Fig. 4.95 Earthquake surface rupture zone offsets river valley, floodplain, road and hills at a site to the northeast of Nanba Town, Pingwu County. Antithetic scarp on river valley and floodplain has blocked the river to form a sag pond. Viewed to northeast

Fig. 4.96 Antithetic fault scarp formed on the riverbed at the outcropped point of the earthquake fault on the southern side of a road at a site to the northeast of Nanba Town, Pingwu County. The scarp is 1–1.5 m high, and the right-slip offset is 2–3 m; two sag ponds were formed close to the fault on the footwall of the scarp. Viewed to southwest

Fig. 4.97 The wheat field, paddy field and corn field were warped and folded to form a 1.0–1.2-m-high scarp at Lizikan Village, Nanba Town, Pingwu County. Viewed to southwest

Fig. 4.98 At Lizikan Village, Nanba Town, Pingwu County, the wheat field and corn field were not only warped and folded to form a scarp, but offset right-laterally 1.3–1.4 m. Viewed to southeast

4.2 Beichuan Surface Rupture Zone

Fig. 4.99 The province highway No. 105 was vertically offset 1.6–1.8 m at a site about 1 km to the south of Shikan Township, Pingwu County. Viewed to NNE

Fig. 4.100 An antithetic fault-related fold scarp occurred on the floodplain of a river valley at Shikan Township, Pingwu County; the trees near the scarp were tilted. Viewed to southwest

Fig. 4.101 Numerous tension-shear cracks developed on the top of a 2.2-m-high fault-related fold scarp to the southwest of Shikan Petrol Station at Shikan Township, Pingwu County; this may indicate the existence of right-slip component. Viewed to northeast

Fig. 4.102 A scarp vertically offset the cement pavement of the Shikan Petrol Station about 3 m at Shikan Township, Pingwu County; the broken cement plates near the scarp are nearly vertical. Viewed to southwest

4.2 Beichuan Surface Rupture Zone

Fig. 4.103 The road near the Shikan Petrol Station at Shikan Township, Pingwu County was not only warped and folded, but its embankment was right-laterally offset about 2.4 m. Viewed to north

Fig. 4.104 An about 0.75-m-high fault-related fold scarp developed on a river shoal near Sujiayuan Village, to the north of Shikan Township, Pingwu County; the scarp extends northeastward along hill slope on the left side of the river valley. Viewed to northeast

Fig. 4.105 Steeply dipping Paleozoic strata outcropped on the hanging wall of the fault-related fold scarp on a river shoal near Sujiayuan Village, to the north of Shikan Township, Pingwu County. Viewed to northeast

Fig. 4.106 A local bedrock normal fault scarp developed on the eastern bank of a river near Sujiayuan Village, to the north of Shikan Township, Pingwu County; the fault plane dips southeast at an angle of 50°, and nearly vertical at individual sites; the bedrock scarp is about 1.7 m high. Viewed to north

4.2 Beichuan Surface Rupture Zone

Fig. 4.107 Slickenside striation with a rake variable in the range of 15°–40° are developed on the bedrock normal fault plane on the eastern bank of a river near Sujiayuan Village, to the north of Shikan Township, Pingwu County; this may indicate the existence of relatively large right-slip component. Viewed to north

Fig. 4.108 On the surface rupture zone in the east of Sujiayuan Village to the north of Shikan Township, Pingwu County, *dark grey* fault zone broken materials were outcropped; the materials are loose and locally powder-like. Viewed to northeast

Fig. 4.109 Terrace-field-tree-like rows of trees to the north of the house of Mr. Liu Changgui's family at Woqian Village, Shiba Township, Qingchuan County, were right-laterally offset by the N55°E-trending tension-shear fracture about 3.4 m. Viewed to southeast

4.2 Beichuan Surface Rupture Zone

Fig. 4.110 At a site near the courtyard of the house of Mr. Liu Changgui's family at Woqian Village, Shiba Township, Qingchuan County, a pine tree was split into two half parts by a NEE-trending tension-shear fracture, displaying an opening width of 36 cm and right-slip offset of 42 cm. Viewed to east

Fig. 4.111 A 12-m-wide and S35°E-trending tensional graben developed on the loess ridge at the 2nd Team of Liangliangshan Gaofeng Village, Shiba Township, Qingchuan County; the graben may represent the eastern tip structure of the Beichuan-Yingxiu surface rupture zone developed along the Central fault. Viewed to south

Fig. 4.112 The Hongguang dammed lake on the dike of the Donghe River, Qingchuan County on the eastern end of the Beichuan-Yingxiu surface rupture zone along the Central fault. Viewed to northwest

4.2 Beichuan Surface Rupture Zone

The Longmen Shan-Qingping secondary rupture zone is located on alpine valley section bounding the high and middle mountains of the Longmen Shan mountain area, having a total length of 62 km. It strikes N(35°±5°)E and is left-stepping to the Hongkou secondary rupture zone, with a discontinuous step-over of 4 km long and 1.5 km wide. It means that a 5 km-long gap of surface rupture exists in between the Longmen Shan-Qingping and Hongkou secondary rupture zones. The Longmen Shan-Qingping secondary rupture zone is a typical pressure ridge with right-slip component. In the vicinity of the Longmen Shan Town, the vertical and horizontal displacement were measured to be 1.0~2.5 m and 0.8~2.0 m (Figs. 4.21, 4.22, 4.23, 4.24 and 4.25), respectively, while the rupture zone is 10~20 m wide. Along the rupture zone, almost all of the buildings were destroyed, while collapse, landslide and mudflow occurred intensively. And especially in the vicinity of Huilonggou village, mountain slide was very severe (Figs. 4.23 and 4.24). It can be postulated from the measured displacement amounts that the Longmen Shan-Qingping secondary rupture zone is also dominated by thrusting with right-slip component, and its vertical offset is greater than the right-slip (Figs. 4.2, 4.21–4.39). The maximum vertical offset was measured to be 3.8 m.

The Beichuan-Nanba secondary rupture zone is the eastern section of the Wenchuan earthquake surface rupture zone. It is right-stepping to the Longmenshan-Qingping secondary rupture zone, making up a 5~6 km wide pull-apart step-over called Gaochuan step-over (Fig. 4.1). As the whole Longmen Shan region is situated under a compressive stress condition, some N80°E-trending compressive fault scarps of several centimeters height with south-facing free face were developed intermittently within the step-over on the Erlangmiao to Baimugou road (31.63100° N, 103.20608 °E). The Beichuan-Nanba secondary rupture zone has a total length of about 135 km and a general strike of N42°±5°E, along which the maximum vertical offset is 6.5±0.5 m (Fig. 4.64) and the maximum right-slip displacement is 4.9 m (Fig. 4.79). Except at Qushan Town (Beichuan County) to Dengjia in the north and from Shikan to the vicinity of Sujiayuan, where the zone becomes a SE-dipping normal fault (Figs. 4.60, 4.61, 4.62, 4.63, 4.64, 4.65, 4.66, 4.67, 4.106 and 4.107), the other sections of the zone is NW-dipping at an angle of about 70° and can be assigned to right-lateral strike-slip reverse fault (Figs. 4.40–4.112), consisting of right-stepping subsidiary ruptures, the right-slip component of which is greater than the vertical slip component (Fig. 4.2). Moreover, along the Beichuan-Nanba secondary rupture zone there exists also a subsidiary step-over, on which abundant phenomena of co-seismic rupturing were well developed. For example, at Leiguzhen Town, which is located at a step-over between the two left-stepping NE-trending strike-slip thrust faults, several NE-, NNE- and NW-striking reverse fault type surface ruptures were developed in addition to numerous back fringe scarps associated with landslide. Their vertical offset is generally in the range of 2.5~3.5 m with a maximum value of 4.5~5 m, and their right-slip displacement is in the range of 1~2.5 m. However, left-lateral displacement was observed also at individual sites. These may indicate that the compressive step-over at Leiguzhen Town is dominated mainly by reverse faulting surface ruptures, but at local section apparent strike-slip component may also occur owing to the variations of fault strike, topography and landform.

102 4 Earthquake Surface Ruptures

According to the measurements of co-seismic displacement and the rake of slickenside striation, the Beichuan surface rupture zone, if taken the 5~6 km-wide Gaochuan step-over as a boundary, can be subdivided into two basic segments of obviously different kinetic features:

(1) The Hongkou-Qingping segment dominated by thrusting with right-slip component. The segment consists of the Hongkou and Longmenshan-Qingping secondary surface rupture zones, stepping to each other, having a total length of about 105 km. The maximum vertical offset is 6.2±0.5 m with an average of 3~4 m and the average right-slip displacement is 1~3 m.
(2) The Beichuan-Nanba segment dominated by right-lateral strike-slip with vertical slip component. The segment has a total length of about 135 km, along which the maximum right-lateral displacement is 4.9±0.2 m with an average of 2~3 m, and the maximum vertical slip is 6.5±0.5 m (After Xu et al., 2009a).

In consideration of the fact that the 5~6 km-wide Gaochuan step-over has separated the Hongkou-Qingping and the Beichuan-Nanba segments of obviously different mechanical properties, and that mechanically it has been able to delay effectively the earthquake rupture propagation of a certain magnitude, the step-over therefore can be considered as a tectonic boundary for the segmentation of secondary surface ruptures (Tim and Ken, 1997; Basile and Brun, 1999; Harris and Day, 1999; Lettis et al., 2002; Tamer et al., 2005). The field observed segmentation of the surface ruptures coincides well with that deduced from the inversion of seismic data. The Wenchuan earthquake resulted from the superimposition of two earthquake rupturing events: the first event that occurred at the western termination of the Hongkou-Qingping segment of the Beichuan fault on the middle section of the Longmen Shan thrust belt was dominated by simple reverse faulting, and when it propagated eastward the second event was triggered to occur on the Beichuan-Nanba segment dominated by right-lateral strike-slip with thrust slip component (Chen et al., 2008a). And that is why the duration time of the Wenchuan earthquake rupturing reached up to 90 s.

4.3 Hanwang Surface Rupture Zone

The Hanwang surface rupture zone along the front-range fault (the Pengguan fault or the Guanxian-Jiangyou fault) of the Longmenshan thrust belt is the second major rupture zone associated with the Wenchuan earthquake. This rupture zone is located ~12 km east of the Beichuan rupture zone, having a total length of 72 km on the surface (Fig. 4.1). It starts from Jian'an Village (31.1650278°N, 103.852694°E) east of Tongji Town, Pengzhou City, in the west, extending N45°±5°E through Bailu Town of Pengzhou City, Bajiao Town of Shifang City and Hanwang Town of Mianzhu City, and ends at Chuanzhu Village (31.6285°N, 104.3720°E), Sangzao Town of Anxian County. Field observation shows that the Hanwang surface rupture zone, in contrast to the Beichuan surface rupture zone, is characterized by pure reverse faulting (Figs. 4.113–4.156). At Shaba Village (31.39811°N, 104.11836°E), Jiulong

4.3 Hanwang Surface Rupture Zone

Fig. 4.113 Brittle fracturing of cement pavement on the west of Bailusi gully, Tongji Town, Pengzhou City. The northwest side of the road was uplifted about 1.55 m in total, associated with 30 cm left-slip offset. Viewed to south

Fig. 4.114 A N40°E-trending fault-related fold scarp about 35-cm-high developed on a corn field on the westernmost end of the Hanwang surface rupture zone in the west of Jian'an Village, Tongji Town, Pengzhou City. The level of corn stems on the hanging-wall was uplifted synchronously with the scarp. Viewed to southwest

Fig. 4.115 Transtensional crack zone developed on the hanging-wall of the N40°E-trending and gently-dipping fault-related fold scarp in the west of Jian'an Village, Tongji Town, Pengzhou City. Viewed to southwest

Fig. 4.116 A N45°E-trending scarp with uplifted northwest wall and descended southeast wall developed at the yard of Mr. Zhu Yang'an at Jian'an Village, Tongji Town, Pengzhou City. The scarp is about 52 cm high. The houses on the scarp were completely destroyed and the trees were tilted. Viewed to northeast

4.3 Hanwang Surface Rupture Zone

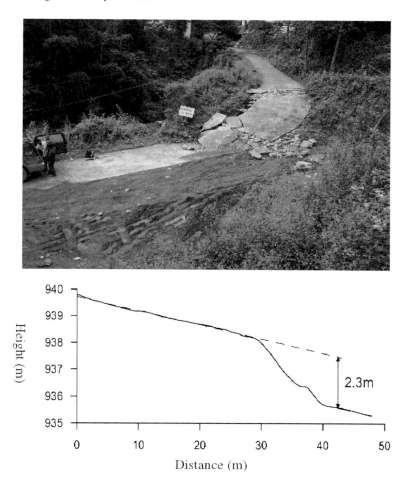

Fig. 4.117 Brittle fracturing of cement pavement within a wide range of 13 m at Bailusi gully, Tongji Town, Pengzhou City, resulting in a 2.3-m-high scarp. Viewed to north

Fig. 4.118 The house of Mr. Ma Liangmin on the surface rupture zone at Shuangyang Village near Bailusi gully, Tongji Town, Pengzhou City, was completely collapsed. Viewed to east

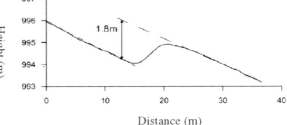

Fig. 4.119 A fault-related fold scarp 15 m wide opposite to the slope of a road was formed at Bailusi gully, Tongji Town, Pengzhou City. The scarp is 1.8 m high, and the trees near the scarp were tilted. Viewed to west

4.3 Hanwang Surface Rupture Zone

Fig. 4.120 A 2.2-m-high fault-related fold scarp cuts obliquely the cement road and peasant's houses and wire poles were completely collapsed at Wangjiaba Village, Bailu Town, Pengzhou City. The rupture zone on cement road is 21 m wide. Viewed to northeast

Fig. 4.121 The tilted corn stems growing on the fault-related fold scarp near the house of Mr. Ma Liangmin at Shuangyang Village near Bailusi gully, Tongji Town, Pengzhou City. The house of Mr. Ma Liangmin in the distance was completely collapsed (Fig. 4.118). Viewed to northeast

Fig. 4.122 An about 1.92 m-high and N50°E-trending fault-related fold scarp on cropland in the north of Wangjiaba Village, Bailu Town, Pengzhou City. Viewed to southwest

Fig. 4.123 A N50°E-trending fault-related fold scarp that cut a ridge between fields with a right-lateral offset of 0.8 m in the north of Wangjiaba Village, Bailu Town, Pengzhou City. Viewed to northwest

4.3 Hanwang Surface Rupture Zone

Fig. 4.124 An about 1.7 m-wide tension crack zone was formed on the hanging-wall of a N50° E-trending fault-related fold scarp in the north of Wangjiaba Village, Bailu Town, Pengzhou City, owing to local extension. Viewed to southwest

Fig. 4.125 A N50°E-trending fault-related fold scarp at Guangou Village, Bailu Town, Pengzhou City Numerous N40°E-trending tension cracks within width of 28 m were developed on the hanging wall of the scarp 2.7 m high. Viewed to east

Fig. 4.126 A fault-related fold scarp at the Central School of Bailu Town, Pengzhou City. The scarp is 1.8 m high and strikes N20°E. The distance between the scarp and the education building on the northwest side is only 3 m, but the building was undamaged, but the structures and buildings across the scarp in distance and close view were completely collapsed or tilted. Viewed to northwest

Fig. 4.127 The new fault scarp about 2.5 m high superimposed on the preexisted fault-related fold scarp in the west of Bailu Town, Pengzhou City. The trees and corn stems at the middle and lower parts of the scarp were tilted, indicating the new scarp location. Viewed to west

4.3 Hanwang Surface Rupture Zone

Fig. 4.128 Waterfall and fault scarp on Laojiehe River and floodplain at Bailu Town. The houses on the extension of the scarp were collapsed. Viewed to northeast by north

Fig. 4.129 Waterfall scarp of Laojiehe River on the Hanwang surface rupture zone at Bailu Town (viewed to north). 3-D scanning of offset topography (on the *left*) and across-scarp tophographic profile (lower *right*) show the scarp is about 2.4 m high

Fig. 4.130 A N15°E-trending and 3.5-m-high fold scarp on the road to Shangshuyuan at Bailu Town, Pengzhou City. Viewed to north

Fig. 4.131 A 1.7-m-high simple pressure ridge obliquely cutting a road at a big col between the Bailu and Bajiao Towns. Viewed to northeast

4.3 Hanwang Surface Rupture Zone

Fig. 4.132 Two sub-parallel brittle breaks occurred on cement pavement of a road and the cement plates were superimposed north of Yinghua Town, Shifang City. The amount of superimposition of cement plates in close view is 1.2 m, and at the site the people standing in front of a car in distance is 0.72 m, indicating crustal shortening of at least 1.9 m. Viewed to northwest

Fig. 4.133 A close-up view of about 1.2 m superimposition of cement plates north of Yinghua Town, Shifang City. Viewed to northwest

Fig. 4.134 A fault-related fold scarp at a large col between southwest Bajiao Town and Bailu Town, Shifang City. The houses on the hanging wall of the scarp were collapsed, and the wire pole standing on the footwall near the scarp bottom was tilted. Viewed to southwest

Fig. 4.135 The deck of a bridge and a single tree trunk on a 1.5-m-high fault-related fold scarp at Shilong Mountain Villa, Xuanlonggou Village in between the Jinhua and Zundao Towns, Mianzhu City were tilted southeastward. The bridge seat was cracked, the houses on the scarp and its hanging wall were collapsed, and the damages of villa buildings on the footwall are slighter. Viewed to southwest

4.3 Hanwang Surface Rupture Zone

Fig. 4.136 The bridge at Shilong Mountain Villa, Xuanlonggou Village in between the Jinhua and Zundao Towns, Mianzhu City, was obliquely cut by a 1.5-m-high fault-related fold scarp. The bridge seat and its deck were tilted and damaged. Viewed to southwest

Fig. 4.137 The villa's buildings on the footwall near the scarp were cracked and warped and the milky white house on the scarp was tilted at Shilong Mountain Villa, Xuanlonggou Village in between the Jinhua and Zundao Towns, Mianzhu City, Viewed to southwest

Fig. 4.138 Brittle rupture occurred on the cement pavement of a road on the hanging wall (away from the scarp) to form a tent-like pressure ridge and the houses nearby were damaged, north of Shilong Mountain Villa, Xuanlonggou Village in between the Jinhua and Zundao Towns, Mianzhu City. Viewed to south

Fig. 4.139 Fault-related fold scarp at Shilong Mountain Villa, Xuanlonggou Village in between the Jinhua and Zundao Towns, Mianzhu City. The scarp is about 1.5 m high, the houses and bridge near the scarp were damaged. The bridge deck was tilted and brittle rupture occurred on the cement road. Viewed to north

4.3 Hanwang Surface Rupture Zone

Fig. 4.140 A 2.0 m-high fault-related fold scarp (in distance) and a pavement suprarhrust scarp in front were developed on a cement road at Longquan Moutain Villa, Qingquan Village, Jiulong Town, Mianzhu City. The scarp is N25°E-trending. The cement plates were superimposed. Its superimposition amount on the hanging wall is about 1.5 m, and that on the footwall is 1.3 m, which shows a total shortening amount of about 2.8 m. Viewed to north

Fig. 4.141 A NNE-trending right-slip pressure ridge cuts obliquely the river channel to form a waterfall on the east side of a dirt road at Shaba Village, Jiulong Town, Mianzhu City. Viewed to northwest

Fig. 4.142 A NNE-trending dextral pressure ridge at Shaba Village, Jiulong Town, Mianzhu City. The coseismic vertical offset measured from the deformed dirt road is about 2.4 m, and the right-lateral slip about 2.9 m. The maximum vertical offset of about 3.5 m for the Hanwang surface rupture zone was obtained at a site eastward not so far from this site. Viewed to west

Fig. 4.143 Waterfall scarp on a man-made canal and surface rupture on the westward extension of the scarp in between the western bank of the Qingshui River and the railway of the Dong Fang Steam Turbine Works at Hanwang Town, Mianzhu City. The Figure shows the tilted embankment and wire poles, as well as the tension cracks on the hanging wall. Viewed to southwest

4.3 Hanwang Surface Rupture Zone

Fig. 4.144 A N55°E-trending and about 17-m-wide fault-related fold scarp cuts the channel bed of the western tributary of the Qingshui River, grass-growing floodplain in the center, a dirt road for transporting sand and stones, as well as the channel bed of the eastern tributary, resulting in a 1.0±0.2 m-high waterfall or scarp. Viewed to north

Fig. 4.145 A close-up view of the cross section of the surface rupture zone, which shows shortening of the cement plates of an old road at the western bank of the Qingshui River, Hanwang Town, Mianzhu City. Besides, the embankment was also broken and offset right-laterally about 55 cm. Viewed to west

Fig. 4.146 A fault-related fold scarp on the cement pavement by the side of a man-made canal in between the western bank of the Qingshui River and the railway of the Dong Fang Steam Turbine Works at Hanwang Town, Mianzhu City. The scarp strikes 50° with a slope angle of 15° and a width of 6–6.5 m. The right-lateral offset is 55±10 cm, and the vertical offset is 1.6–2 m. Viewed to northwest

Fig. 4.147 A N50°E-trending scarp on the western bank of the Qingshui River at Hanwang Town. The scarp offsets vertically the corn field on the T1 terrace, and the broken cement pavement of an old road beneath the field displays a shortening amount of about 1.7 m. Viewed to west

4.3 Hanwang Surface Rupture Zone

Fig. 4.148 A simple pressure ridge occurred on a garbage heap on the T1 terrace at the eastern bank of the Qingshui River, Hanwang Town, Mianzhu City. The ground is locally nearly upright, and the scarp is 1.3±0.3 m high. Viewed to northwest

Fig. 4.149 The herbs growing on the simple pressure ridge on the garbage heap on the T1 terrace at the eastern bank of the Qingshui River, Hanwang Town, Mianzhu City, were tilted with the coseismic deformation of the ground. The phenomenon indicates that the northwest wall of the scarp was thrusted over the southeast wall. Viewed to northwest

Fig. 4.150 Two imbricated N45°–50°E-trending simple pressure ridges occurred on the road from Mianzhu to Maoxian at the eastern bank of the Qingshui River, Hanwang Town, Mianzhu City. The two ridges are 80 m apart and the central line of the road is right-laterally offset 47 cm. Viewed to south

Fig. 4.151 The cement plates of western dyke of the river at Quanxin Village east of Hanwang Town, Mianzhu City, are fractured to form tent-like structure superimposed on the fault-related fold scarp. Viewed to northeast

4.3 Hanwang Surface Rupture Zone

Fig. 4.152 Cement plates were superimposed each other on an abandoned road at the western side of the road from Mianzhu to Maoxian (31°27′41.6″N, 104°9′59.0″E) at the eastern bank of the Qingshui River, Hanwang Town, Mianzhu City. The superimposition amount or the crustal shortening is at least 1.2±0.1 m. Viewed to east

Fig. 4.153 Broken cement pavement and about 1-m-high fault-related fold scarp on the western dyke of the river at Quanxin Village, east of Hanwang Town, Mianzhu City. The sideline of the dyke shows that no horizontal offset occurred. Viewed to northeast

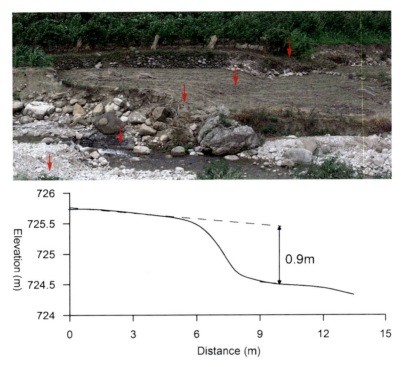

Fig. 4.154 A 94-cm-high fault-related fold scarp on the T1 terrace of the river at Quanxin Village east of Hanwang Town, Mianzhu City. Viewed to northwest by north

Town, Mianzhu City, the maximum vertical offset is measured to be 3.5±0.2 m across a NE-trending fault scarp. Westward from this site, it was observed that the surface rupture dissects a river bed to form a waterfall (Fig. 4.141), where the rupture becomes N15°E-trending and cuts obliquely a nearly north-south-extending dirt road, resulting in a 2.4 m-high pressure ridge with 2.9 m left-slip (Fig. 4.142). This may correspond to about 5 m crustal shortening across NW-SE direction produced by the maximum vertical displacement, on the assumption that the whole Pengguan fault is N45°E-trending and NW-dipping at an angle of about 35°. Therefore, the left-slip offset observed on a N15°E -trending scarp at Shaba Village can be considered as the result of local northward turning of fault strike. Northeastward or southwestward from Shaba Village, the surface rupture turns back to N45° E-trending again, where right-lateral slip of less than 1 m can be observed. Westward from Shaba Village to Bailu Town, the vertical offset remains to be about 2 m (Fig. 4.2). At Bailu Town, a N45°E-trending rupture zone cuts the Laojie River to form a waterfall with vertical offset of up to 2.4 m (Figs. 4.128 and 4.129). Further westward, the rupture zone passes through the space between two 20 m-apart education buildings of the Bailu Middle School, resulting in a N30°E-trending scarp 1.8 m-high (Fig. 4.126). Surprisingly, the education building on the footwall of the

4.3 Hanwang Surface Rupture Zone

Fig. 4.155 A nearly 20-cm-high fault scarp and the tilted trees in the vicinity of Chuanzhu Village (31.62850°N, 104.37200°E), Sangzao Town, Anxian County. Viewed to west

Fig. 4.156 Landscape of dammed earthquake lake on the hanging wall of the Hanwang surface rupture zone on the road from Suishui to Gaochuan Township, Anxian County. Viewed to north

scarp built in 1998 fortified according to intensity VII seismic design was fractured but not collapsed, and the education building that was built in 2005 on the hanging-wall just 3 m away from the scarp was also scarcely cracked on its building walls. In contrast, the other buildings on the scarp were totally collapsed during the earth-quake (Yu et al., 2009). This may indicate that away from the seismic fault (scarp) may mitigate effectively the earthquake hazards.

4.4 Xiaoyudong Surface Rupture Zone

The Xiaoyudong surface rupture zone is a N35°±5°W-trending secondary surface rupture produced by the Wenchuan earthquake. It is located between the western end of the Hanwang rupture zone and the step-over of two stepped Hongkou and Longmenshan-Qingping secondary rupture zones of the Beichuan fault, having a length of about 7 km (Fig. 4.1). The rupture zone is nearly perpendicular to the two main rupture zones and characterized by reverse left-lateral strike-slip (Figs. 4.157–4.180). Taken Xiaoyudong Village, Pengzhou City, as a center, the rupture zone deviates from NW-trending to nearly north-south-trending, and obliquely links with the NE-trending rupture zones on its both ends. Several tension cracks or en-echelon transtensional cracks were well developed on top of the scarp (Figs. 4.157, 4.159, 4.160 and 4.180). The rupture or deformation zone of the scarp is 5–20 m wide and is characterized by the uplift on its southwest wall relative to its northeast wall mak-ing up a slope angle of 30°–35°. The fault scarp crosscuts roads, field ridges, and enclosing walls. It cuts the river valley, terrace and high flood plain of the Baishuihe River to form fault scarp and waterfall landscapes, and also obliquely cuts the east-ern Xiaoyudong bridge seat (Figs. 4.165 and 4.166), causing the breaking of the bridge on its several sections. The structures and buildings on the scarp were seri-ously damaged (Figs. 4.162, 4.166, 4.170, 4.174, 4.175 and 4.179), while the crops and trees were tilted to form "drunkard woods" (Figs. 4.174 and 4.175). Simple pressure ridges and fault-related fold scarps were formed on two cement roads at Xiaoyudong town. (Figs. 4.167–4.180). At the observation site on the northern road (31.19419°N, 103.75450°E), the scarp was measured by total station to be 1.1 m high, and its left-slip offset to be 2.3 m (Fig. 4.169). The road and enclos-ing walls at Jianjiang Chemical Plant were left-laterally offset synchronously 1.2 m (Fig. 4.168). At the 2nd Group of Luoyang Village (31.19670°N, 103.75194°E), the maximum left-lateral offset was measured to be 3.5 m, while the vertical offset only 1.9 m (Fig. 4.173). Northward from this site, the rupture zone becomes to be nearly north-south- or NNE-trending. On the lower terrace at the northern end of the zone (31.19670°N, 103.75194°E), a 3.4 m high simple pressure ridge was formed on a sand-and-stone road (Fig. 4.176) and the ground of a coal plant, reflecting the max-imum vertical offset on the Xiaoyudong rupture zone without significant strike-slip component.

4.4 Xiaoyudong Surface Rupture Zone

Fig. 4.157 The main scarp and associated tension-cracks of the large scale landslide at a site between Cifeng, Pengzhou City and Xiang'e, Dujiangyan City. Viewed to northeast

Fig. 4.158 Southeastward tilting of paddy field at the 9th Group of Honghuo Village, Xiang'e Township, Dujiangyan City. The water depth is about 30 cm, and the field ridge on water-stored side was bent. This figure shows that the farmers are leveling the tilted paddy field. Viewed to northeast

Fig. 4.159 Large scale landslide at a site between Cifeng, Pengzhou City and Xiang'e, Dujiangyan City. Numerous tension cracks were developed within the landslide. Viewed to northwest

Fig. 4.160 A bent road and tension cracks were formed at the 9th Group of Honghuo Village, Xiang'e Township, Dujiangyan City. Viewed to north

4.4 Xiaoyudong Surface Rupture Zone

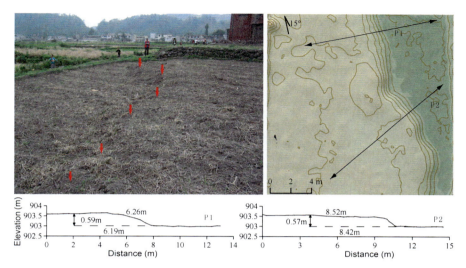

Fig. 4.161 A nearly north-south-trending fault-related fold scarp at Caoba Village, Tongji Town, Pengzhou City. The scarp is about 58 cm high. Viewed to south

Fig. 4.162 The NNW-trending fault-related fold scarp left-laterally offsets a paddy field at Caoba Village, Tongji Town, Pengzhou City. Viewed to west

Fig. 4.163 The NW-trending fault-related fold scarp left-laterally offsets the field ridge 30 cm or more at Caoba Village, Tongji Town, Pengzhou City. Viewed to northwest

Fig. 4.164 The embankment of the Jianjiang River was left-laterally offset 30 cm. Viewed to east

4.4 Xiaoyudong Surface Rupture Zone

Fig. 4.165 The Xiaoyudong surface rupture zone obliquely cuts the river valley of the Baishuihe River and the eastern seat of the Xiaoyudong Bridge. The bridge on hanging wall was broken into several sections. Viewed to north

Fig. 4.166 The Xiaoyudong surface rupture zone obliquely cuts the eastern seat of the Xiaoyudong Bridge, causing fracturing of the oblique beam and the bending of the reinforced bars of the bridge. Viewed to north

Fig. 4.167 Sinistral pressure ridge and its left-lateral offset on the terrace of the Jianjiang River. Viewed to south

Fig. 4.168 The sinistral pressure ridge cutting through the enclosing walls, road and cropland at the Jianjiang Chemical Plant. The vertical offset was measured to be 1.2 m with a left-slip component. Viewed to southwest

4.4 Xiaoyudong Surface Rupture Zone

Fig. 4.169 A 1.47-m-high and NW-trending sinistral pressure ridge on a road to the north of Xiaoyudong Town. The left-slip offset is 2.3 m, and the cement pavement of the road near the scarp was broken. Viewed to southwest

Fig. 4.170 A NW-trending fault-related fold scarp cuts the road in Xiaoyudong Town to form an about 1.09-m-high scarp. The houses on the scarp were completely collapsed. Viewed to northwest

Fig. 4.171 The cropland and its filed ridge at Luoyang Village, Xiaoyudong Town, were deformed during the earthquake to form a sinistral pressure ridge. The vertical offset of the field ridge is 1.2 m and the left-lateral offset is 1.5 m. Viewed to southwest

Fig. 4.172 The sinistral pressure ridge cuts the yard in front of a storied building of Mr. Ke Meihong's family and the cropland at Luoyang Village, Xiaoyudong Town. Viewed to southwest

4.4 Xiaoyudong Surface Rupture Zone

Fig. 4.173 The maximum left-lateral offset of the Xiaoyudong surface rupture zone was measured near the high-tension wire line at the 2nd Group of the Luoyang Village, Xiaoyudong Town, to be 3.5 m (the earth bank between the two standing men), and the vertical offset was measured to be 1.9 m. Viewed to southwest

Fig. 4.174 Landscapes of a fault-related fold scarp on upper alluvial-pluvial fan at the 2nd Group of the Luoyang Village, Xiaoyudong Town. From top to front edge of the scarp, the trees and wire poles change gradually from vertically-standing, tilting to lying down. On the footwall the plants have become a typical "drunkard woods". Viewed to north northwest

Fig. 4.175 A close-up view of the so-called "drunkard woods". Viewed to northwest by north

Fig. 4.176 A hanging-wall collapsed fault scarp on a road of the floodplain, eastern bank of a river to the northeast of Luoyang Village, Xiaoyudong Town. The scarp is NEE-striking and about 3.4 m high. Viewed to north

4.4 Xiaoyudong Surface Rupture Zone

Fig. 4.177 The offset landform of the earthquake surface rupture zone showing scarps and waterfall on T1 and T2 terraces at the site to the northeast of Luoyang Village, Xiaoyudong Town. Viewed to northeast by north

Fig. 4.178 The waterfall that was formed during the earthquake to the northeast of Luoyang Village, Xiaoyudong Town. Viewed to north

Fig. 4.179 The houses near the Xiaoyudong surface rupture zone were seriously damaged. Viewed to north

Fig. 4.180 A hanging-wall collapsed fault scarp on eastern side of a road at the floodplain on the eastern bank of a river to the northeast of Luoyang Village, Xiaoyudong Town. Tension cracks were well developed on its hanging wall. Viewed to northeast

4.5 Appendix 4.1: Co-seismic Offsets Along the Beichuan Rupture Zone

The surface ruptures associated with the Wenchuan earthquake were well measured during the scientific investigation. The co-seismic offsets along the Beichuan rupture zone are listed below (See Figs. 4.1 and 4.2 for detailed locations).

No	Lon E	Lat N	Slip type	Strike slip (cm)	Vertical slip (cm)	Offset features
1	103.4828611	31.0613889	R		100	Terrace
2	103.4828611	31.0613889	R		100	Plantation
3	103.4828611	31.0614167	R		100	Highway surface
4	103.4855278	31.0631111	R		150	Plantation
5	103.4855278	31.0631111	R		150	Plantation
6	103.4855833	31.0630833	R		173	Highway surface
7	103.486306	31.064972	R		110	IV Terrace of Min River
8	103.487278	31.064028	RL, R	60	100	IV Terrace of Min River
9	103.487306	31.063806	R		80	IV Terrace of Min River
10	103.4896667	31.0652778	R		230	Highway surface, terrace of Min River and waterfall
11	103.4897222	31.0653056	R		200	Highway surface, young terrace of Min River
12	103.4897222	31.0653056	R		200	Plantation
13	103.489861	31.065222	RL, R	66	80	Road
14	103.5654722	31.0832500	R		150	Valley plantation
15	103.5960556	31.0740556	RL, R	100	170	Valley plantation, field ridge
16	103.6118333	31.0865556	RL	400	?	Road and terrace riser
17	103.6126944	31.0875833	RL	160	?	Road and terrace riser
18	103.6148889	31.0894722	R		500	Young terrace and road
19	103.6149167	31.0892500	R		400	Young terrace and road
20	103.6149167	31.0892500	R		400	Terrace
21	103.6157778	31.0899444	RL, R	280	480	Road margin and surface
22	103.6158056	31.0899444	R, RL	450	270	Concrete road margin and its surface
23	103.6161111	31.0902500	R		370	Terrace
24	103.6169722	31.0912222	R		100	Terrace

No	Lon E	Lat N	Slip type	Strike slip (cm)	Vertical slip (cm)	Offset features
25	103.6169722	31.0912222	R		45	Terrace
26	103.6169722	31.0912222	R		77	Terrace
27	103.6215833	31.1004167	R		330	Terrace
28	103.6222500	31.1047200	R		620	Concrete road surface and courtyard
29	103.653056	31.1192500	RL, R	280	?	Plantation
30	103.6553056	31.1192500	RL, R	280		Road margin
31	103.6553056	31.1193333	RL, R	280	250	Road margin and terrace riser
32	103.6564722	31.1197222	RL, R	430	120	Plantation
33	103.6618611	31.1231389	RL, R	400	300	Terrace and its riser
34	103.6642778	31.1252500	RL, R	120	50	Terrace and its riser
35	103.6707778	31.1253056	RL, R	75	260	Terrace and its riser
36	103.6918889	31.1452222	R		400	Terrace
37	103.6918889	31.1452222	R		400	Plantation
38	103.6920000	31.1453611	R		460	High terrace and road
39	103.8183056	31.2855000	R	220	250	Road and its margin
40	103.8183056	31.2855000	R		200	
41	103.8183056	31.2855000	R	200	250	Road and its margin
42	103.8303889	31.2857222	RL, R	80	100	Road and its margin
43	103.8484722	31.2941944	R		500	Courtyard
44	103.9950600	31.4534600	R		340	Alluvial fan
45	104.0709167	31.5339722	R		300	Road
46	104.1039167	31.5654722	R		300	Terrace
47	104.1094000	31.5699900	R	55	370	Terrace
48	104.1096667	31.5703333	RL, R	150	380	Highway and its margin
49	104.1100000	31.5680278	R		350	Terrace
50	104.1132778	31.5724722	R		350	Woodland
51	104.1290278	31.5850278	R		350	Terrace
52	104.1305556	31.5868333	R		320	Terrace
53	104.1312778	31.5873889	R		325	Terrace
54	104.154028	31.618722	RL, R	470?		Not sure
55	104.1544444	31.6188611	RL, R		320	Plantation
56	104.1545833	31.6188888	RL, R		205	Plantation
57	104.1731700	31.6286400	RL, R		315	Channel and terrace
58	104.1743211	31.6292731	RL, R		259	Terrace
59	104.1744444	31.6292777	RL, R	160	319	Terrace and bridge
60	104.1754722	31.6296666	RL, R		280	House and its courtyard

4.5 Appendix 4.1: Co-seismic Offsets Along the Beichuan Rupture Zone

No	Lon E	Lat N	Slip type	Strike slip (cm)	Vertical slip (cm)	Offset features
61	104.1759444	31.6297222	RL, R	165	135	Woodland, ridge of field
62	104.1767222	31.6298888	RL, R		220	Road
63	104.2060833	31.6310000	RL, R		16	Road
64	104.2811111	31.6470556	RL, R	430	520	Rock margin and scarp
65	104.2811111	31.6470556	RL, R	430	520	Bedrock
66	104.3820000	31.7248000	R		150	?
67	104.3821600	31.7248600	R		150	?
68	104.3822100	31.7250000	RL, R	120	?	?
69	104.3822700	31.7249200	RL, R	120	150	
70	104.3822700	31.7249200	R		150	?
71	104.3823400	31.7249200	RL, R	120	?	?
72	104.38324	31.7250000	RL, R		160	
73	104.4084100	31.7446000	R		180	Terrace
74	104.4199722	31.7803611	R		500	Terrace
75	104.42002	31.78103	R		100	
76	104.4201944	31.7803056	R		419	Terrace
77	104.4202	31.78126	R		100	
78	104.4202100	31.7804300	R		340	Young terrace
79	104.4202600	31.7812900	R		100	Young terrace
80	104.4202700	31.7805700	R		340	Young terrace
81	104.4203000	31.7807700	R		340	Young terrace
82	104.4203000	31.7813300	R		100	Young terrace
83	104.4203056	31.7805278	R		300	Young terrace
84	104.4203100	31.7802400	R		340	Terrace
85	104.4203100	31.7813300	R		100	Young terrace
86	104.4203700	31.7798300	R		310	Terrace
87	104.4203800	31.7807200	R		230	Terrace
88	104.4203800	31.7808600	R		230	Terrace
89	104.4203889	31.7806944	R		314	Terrace
90	104.4203900	31.7809100	R		230	Terrace
91	104.4203900	31.7809600	R		230	Terrace
92	104.4204000	31.7801100	R		340	Terrace
93	104.4204100	31.7801000	R		340	Terrace
94	104.4204100	31.7809400	R		230	Terrace
95	104.4204700	31.7816500	R		100	Terrace
96	104.4204722	31.7810833	R	100		Terrace
97	104.4204800	31.7800600	R		340	Terrace
98	104.4205000	31.7810800	R		230	Terrace
99	104.4205300	31.7811900	R		230	Terrace
100	104.42054	31.78148	R		100	?
101	104.4206902	31.7812700	R		190	Terrace
102	104.4206000	31.7812000	R		230	Terrace
103	104.4206700	31.7813100	R		317	Terrace
104	104.4207000	31.7812700	R		340	Waterfall
105	104.4220000	31.7792900	R		95	Plantation
106	104.4220278	31.7792778	R		40	Plantation (secondary)

No	Lon E	Lat N	Slip type	Strike slip (cm)	Vertical slip (cm)	Offset features
107	104.4221667	31.7793333	RL, R	20	95	Plantation (main)
108	104.4225000	31.7792900	RL, R	120	211	Plantation
109	104.4225300	31.7793100	RL, R		385	Plantation
110	104.4225556	31.7793056	RL, R	200	390	Ground work for hogpen
111	104.422556	31.779306	RL, R	130		Plantation
112	104.4226900	31.7789900	RL, R	80	307	Plantation
113	104.4262222	31.7689722	RL, R	140	150	Plantation
114	104.42635	31.76972	RL, R		50	?
115	104.4263500	31.7697200	R		50	?
116	104.4264600	31.7754100	R		520	Plantation
117	104.4265	31.76989	RL, R		70	
118	104.4265000	31.7698900	R		70	?
119	104.42666	31.76994	RL, R		50	
120	104.4266600	31.7699400	R		50	?
121	104.4271389	31.7988056	R	210	250	Plantation
122	104.4272000	31.7989000	RL, R	246	254	Plantation
123	104.4273900	31.7705300	RL, R	40		Plantation
124	104.4275000	31.7994000	RL, R	220	360	Road
125	104.4277800	31.7995000	RL, R	228	380	
126	104.4278000	31.7995000	RL, R	228	380	Plantation, ridge of field
127	104.4281000	31.8000000	RL, R		450	
128	104.4281000	31.8000000	RL, R		450	Plantation
129	104.4286000	31.8014000	RL, R	143	160	Road
130	104.4316200	31.8057800	RL, R		430	Road and plantation
131	104.4317700	31.8060000	RL, R		390	Road
132	104.4463000	31.8151000	R		250	Road
133	104.4469400	31.8154900	R		125	Road
134	104.4475700	31.8152000	R		25	Road
135	104.4476300	31.8156100	R		212	Gully
136	104.4476500	31.8157100	R		424	Hillside deposit
137	104.4476944	31.8157222	R		430	Hillside deposit
138	104.4568889	31.8289444	R	214	305	Plantation
139	104.4568900	31.8289400	R, RL	240	310	Highway and its margin
140	104.4619444	31.8331111	N, RL		220	
141	104.4620556	31.8331389	N, RL		70	
142	104.4622500	31.8332778	N, RL	370	65	Road
143	104.4629167	31.8335833	N, RL	250	285	Gully bottom
144	104.4643888	31.8349166	N, RL		400	Lowland on the hillside and bedrock scarp
145	104.4681500	31.8383333	N, RL		250	
146	104.4681500	31.8383333	N, RL	380	650	Lowland on the hillside and bedrock scarp

4.5 Appendix 4.1: Co-seismic Offsets Along the Beichuan Rupture Zone

No	Lon E	Lat N	Slip type	Strike slip (cm)	Vertical slip (cm)	Offset features
147	104.4695000	31.8394167	N, RL	270	490	Road
148	104.4698055	31.8396944	N, RL	200	570	Lowland on the hillside and bedrock scarp
149	104.4711389	31.8405555	N, RL		300	
150	104.4715556	31.8410000	N, RL		400	Planted land
151	104.4717222	31.8410556	N, RL	415	290	Man-made terrace lands and ridges
152	104.4721944	31.8413889	N, RL		70	
153	104.4723611	31.8415556	N, RL		400	
154	104.4729444	31.8420278	N, RL	400	600	
155	104.4740556	31.8428611	N, RL	330	600	Pavement
156	104.4742222	31.8430000	N, RL		400	
157	104.4745556	31.8435278	N, RL		450	
158	104.4753056	31.8440556	N, RL	240	200	
159	104.4758889	31.8444444	N, RL	230		Gully bottom
160	104.4763333	31.8449444	N, RL		90	
161	104.4776667	31.8458889	N, RL		150	
162	104.4784444	31.8463096	N, RL		150	
163	104.4976944	31.8563055	N, RL		320	Lowland on the hillside and bedrock scarp
164	104.4989444	31.8568611	N, RL	388	300	Man-made terrace lands and ridges
165	104.4990000	31.8568055	N, RL	430	300	Man-made terrace lands and ridges
166	104.4990000	31.8569444	N, RL	430	280	Man-made terrace lands and ridges
167	104.4990277	31.8570000	N, RL	435	290	Man-made terrace lands and ridges
168	104.4990277	31.8570277	N, RL	422	240	Man-made terrace lands and ridges
169	104.5039444	31.8615277	R		410	Young terrace
170	104.532444	31.873722	RL, R	130	380	
171	104.5335000	31.874389	RL, R	120	360	
172	104.5339167	31.8741389	RL, R	140	270	Plantation, ridge of field
173	104.5339167	31.8741389	RL, R	140	370	Plantation, ridge of field
174	104.5339167	31.8741389	RL, R		240	Plantation
175	104.533917	31.874139	RL, R	140	270	
176	104.5792778	31.9238333	RL, R	220	200	Plantation, ridge of field

No	Lon E	Lat N	Slip type	Strike slip (cm)	Vertical slip (cm)	Offset features
177	104.5792778	31.9238333	RL, R	220	200	Plantation, ridge of field
178	104.5792778	31.9238333	RL, R	345	245	Plantation, ridge of field
179	104.5792778	31.9238333	RL, R	260	?	Road
180	104.579278	31.923833	RL, R	345	245	
181	104.619722	31.993611	RL, R	140	190	
182	104.6197222	31.9936111	RL, R	140	190	Plantation, ridge of field
183	104.619917	31.993944	R		170	
184	104.620194	31.994389	RL, R		200	
185	104.6201944	31.9943889	RL, R		200	Terrace
186	104.621250	31.994972	R		260	
187	104.6217500	31.995583	RL, R	170	230	
188	104.621833	31.995556	RL, R		280	
189	104.6218333	31.9955556	RL, R		280	Terrace
190	104.621944	31.995556	RL, R	240	238	
191	104.6219444	31.9955556	RL, R	240	238	Road and its margin
192	104.622	31.995472	RL, R	235	234	
193	104.6220000	31.9954722	RL, R	235	234	Plantation, ridge of field
194	104.623611	31.996528	RL, R	270	285	
195	104.6236111	31.9965278	RL, R	270	?	?
196	104.6236111	31.9965278	RL, R	235	285	Road
197	104.6456667	32.0160278	R	250	?	Channel and waterfall
198	104.6456667	32.0160278	R		250	Plantation
199	104.645667	32.016028	RL, R	250	250	
200	104.6476111	32.0225278	R	250	250	Plantation, ridge of field
201	104.6476111	32.0225278	R		250	Plantation
202	104.6772500	32.0540278	RL, R	490	150	Terrace, road and its margin
203	104.6773333	32.0540833	RL, R	465	150	Terrace, road and its margin
204	104.6776944	32.0545000	RL, R	440	240	Terrace, road and its margin
205	104.6780000	32.0547500	R	540?	330	Terrace and land ridge
206	104.6782777	32.0550277	RL, R	470	360	Terrace and road
207	104.6783611	32.0549722	R	400	400	Terrace, road margin
208	104.6785556	32.0552500	R	380	400	Terrace, road margin
209	104.6785833	32.0543889	RL, R	400	370	Terrace, road margin
210	104.6790278	32.0549444	RL, R	380		Road margin
211	104.6791389	32.0557778	RL, R	370	400	Terrace, road margin

4.5 Appendix 4.1: Co-seismic Offsets Along the Beichuan Rupture Zone

No	Lon E	Lat N	Slip type	Strike slip (cm)	Vertical slip (cm)	Offset features
212	104.6793889	32.0559444	R		370	Terrace
213	104.6868055	32.0621388	R		253	Terrace
214	104.6875000	32.0619444	RL, R	190	200	Hillside
215	104.6875000	32.0623889	R		230	Waterfall, channel
216	104.6875556	32.0623611	RL, R	190	200	Terrace
217	104.687611	32.062556	RL, R	220	170	
218	104.6886944	32.0636111	RL, R	185	200	Terrace
219	104.688722	32.063500	RL, R	220	210	
220	104.6887778	32.0635556	RL, R	185	250	High terrace, road margin
221	104.6888889	32.0633333	RL, R	185	200	High terrace, road margin
222	104.690139	32.064306	R		110	
223	104.690639	32.065028	R		180	
224	104.838083	32.201417	RL, R	240	170	
225	104.838444	32.201611	R		150	
226	104.839444	32.202556	RL, R	180	150	
227	104.8406111	32.2035556	RL, R	250	150	Terrace, road margin
228	104.8433611	32.2060833	RL, R	160	120	Rows of corn field, road
229	104.8433889	32.2061389	RL, R	160	120	Rows of corn field, road
230	104.8438611	32.2064722	RL, R		120	Rows of corn field
231	104.8438889	32.2065278	RL, R	240	100	Rows of corn field, road
232	104.8438889	32.2065278	RL, R	240	90	Rows of corn field, road
233	104.8490833	32.2100556	RL, R	130	85	Rows of corn field, road
234	104.8493333	32.2102222	RL, R	140	85	Rows of corn field, road
235	104.859778	32.218472	R		100	
236	104.859861	32.218222	RL, R	120	70	
237	104.8764722	32.2352778	RL, R	210	190	Highway
238	104.8778889	32.2369444	RL, R		180	Land
239	104.8793055	32.2381111	RL, R		220	Paddy field
240	104.8799166	32.2386111	RL, R	305	240	Gasoline station, terrace riser, highway
241	104.887139	32.245444	R		170	
242	104.9451666	32.2856944	R		75	Young terrace
243	104.9461944	32.2869722	N, RL		170	Corn field and bedrock scarp
244	104.9466388	32.2871388	N, RL	220		Ridge of field
245	104.9468889	32.2873611	N, RL	140	170	Gully and its eastern bank

No	Lon E	Lat N	Slip type	Strike slip (cm)	Vertical slip (cm)	Offset features
246	104.9472778	32.2877778	R	350	107	Road, corn field
247	104.9480556	32.2883056	N, RL		90	Courtyard
248	104.9733889	32.3116389	R	280	0	Tree line and land
249	104.9737222	32.3117778	RL, R	340	52	Tree line and land
250	105.0317778	32.3461389	R		53	Courtyard
251	105.0339167	32.3471944	N	0	0	High land (graben depth)
252	105.108477	32.403803		0	0	Hill slope

Note: R: Reverse vertical offset; RL: Right-lateral offset; N: Normal vertical offset.
Question marks represent unclear features.

4.6 Appendix 4.2: Co-seismic Offsets Along the Hanwang Rupture Zone

The surface ruptures associated with the Wenchuan earthquake were well measured during the scientific investigation. The co-seismic offsets along the Hanwang rupture zone are listed below (See Figs. 4.1 and 4.2 for detailed locations).

No	Lon E	Lat N	Slip type	Strike slip (cm)	Vertical slip (cm)	Offset features
1	103.8533889	31.1655278	R		35	Corn field
2	103.8544722	31.1665000	R		60	Small road
3	103.8550278	31.1668889	R		52	Courtyard of Mr. Zhu Yang'an
4	103.8595556	31.1699722	R		70	Road
5	103.8634444	31.1729722	R		230	Road
6	103.8649722	31.1740000	R		180	Plantation
7	103.8984167	31.2016389	R		220	Plantation
8	103.8993889	31.2026111	RL, R	80	190	Gully, small road
9	103.9011111	31.2038056	R		270	Plantation
10	103.9013611	31.2040000	R		275	Young terrace, waterfall
11	103.9124722	31.2114722	R		200	School yard
12	103.9132500	31.2117500	R		150	Hillside deposit
13	103.9141944	31.2129722	RL, R	96	240	Young terrace, waterfall
14	103.9142222	31.2129167	LL, R	100	240	Young terrace, waterfall

4.6 Appendix 4.2: Co-seismic Offsets Along the Hanwang Rupture Zone

No	Lon E	Lat N	Slip type	Strike slip (cm)	Vertical slip (cm)	Offset features
15	103.9155278	31.2137778	R		230	Highway
16	103.9160278	31.2143611	RL, R	60		Narrow road, plantation
17	103.9208056	31.2175833	R	:	100	Plantation ?
18	103.9313056	31.2238611	R		45	Plantation
19	103.9334167	31.2250278	R		138	Plantation
20	103.9350833	31.2275556	RL, R	50	105	Plantation
21	103.9883333	31.2927222	R		55	Gully, waterfall
22	103.9892778	31.2983056	R		50	Plantation
23	103.9930833	31.2960833	R		60	Corn field
24	104.0444167	31.3451111	RL, R	18	?	Plantation, ridge of field
25	104.0453889	31.3452500	R		110	Corn field
26	104.0819444	31.3690000	R		150	Courtyard, bridge
27	104.1182500	31.3995278	R		130	Plantation
28	104.1182778	31.3999167	R		350	Alluvial fan
29	104.1183600	31.3981100	LL, R	290	240	Plantation
30	104.1183056	31.3980278	LL, R	290	200	Road and its margin
31	104.1183889	31.3981111	LL, R	280	200	Road margin, waterfall
32	104.1232500	31.3969722	R		110	Plantation
33	104.1279722	31.4128889	R		200	Plantation
34	104.1529722	31.4370833	RL, R	25	?	Hillside
35	104.1589167	31.4607778	R		80	Plantation
36	104.1626389	31.4598611	RL	52	55	Plantation
37	104.1632778	31.4601389	RL	55	134	Terrace and highway margin
38	104.1643611	31.4605833	R		100	Road and waterfall
39	104.1650556	31.4611667	R		80	Terrace, road
40	104.1655556	31.4615278	RL, R	47	75	Highway and its central line
41	104.2035000	31.4781944	R		90	Young terrace
42	104.2038611	31.4779444	R		30	Terrace
43	104.2321667	31.5123056	R		46	Road
44	104.3304700	31.6049000	R		20	Hillside
45	104.3498333	31.6130556	R		25	Young terrace
46	104.3718611	31.6284167	R		18	Plantation
47	104.3720000	31.6285000	R		18	Woodland

Note: R: Reverse vertical offset; RL: Right-lateral offset; LL: left-lateral offset.

4.7 Appendix 4.3: Co-seismic Offsets Along the Xiaoyudong Rupture Zone

The surface ruptures associated with the Wenchuan earthquake were well measured during the scientific investigation. The co-seismic offsets along the Xiaoyudong rupture zone are listed below (See Figs. 4.1 and 4.2 for detailed locations).

No	Lon E	Lat N	Slip type	Strike slip (cm)	Vertical slip (cm)	Offset features
1	103.718	31.1085	R		30	High terrace
2	103.7501	31.19933	RL, R	160	315	Flat land
3	103.7502	31.20056	R		350	Terrace and road
4	103.7518	31.19683	R	93	200	Road
5	103.7519	31.19669	R	350	190	Small gully
6	103.7527	31.19575	R	296	190	Road, ridge of field
7	103.7529	31.19536	R	240	?	Plantation, ridge of field
8	103.7539	31.19494	L, R	100	200	Plantation, ridge of field
9	103.7539	31.19497	L, R	192	170	Plantation, ridge of field
10	103.7545	31.19419	L, R	230	147	Highway and its margin
11	103.7545	31.19417	L, R	230	130	Plantation
12	103.7548	31.19406	L, R	100	120	Plantation
13	103.7548	31.19406	L, R	180	150	Plantation
14	103.7566	31.19247	L, R	30	40	Plantation
15	103.7575	31.19261	L, R	30	109	Highway and its margin
16	103.7585	31.12567	RL, R	45	20	Plantation
17	103.7585	31.12511	RL	10	?	Plantation
18	103.7593	31.19075	RL	120	?	Plantation
19	103.7595	31.12606	RL, R	15	6	Terrace and riser
20	103.7621	31.18889	L, R	300	120	Terrace and riser
21	103.7637	31.18828	L, R	130	150	Terrace and riser
22	103.7916	31.16069	L, R		40	Terrace and riser
23	103.7921	31.16267	L, R	50	40	Terrace and riser
24	103.7922	31.16411	L, R	38	30	Plantation, ridge of field
25	103.7923	31.16111	R		100	Plantation, ridge of field
26	103.7927	31.16447	L, R		30	Plantation, ridge of field
27	103.7928	31.16519	L, R	30	?	Plantation, ridge of field

Note: R: Reverse vertical offset; RL: Right-lateral offset; L: left-lateral offset.

Chapter 5
Earthquake Disasters and Damage Features of Structures and Buildings

The Wenchuan earthquake has not only produced surface ruptures with complicated structures, but also caused tremendous earthquake disasters.

5.1 Seismic Intensity Distribution

In seismology, a scale of seismic intensity is a way of measuring or rating the effects of an earthquake at different sites. Numerous intensity scales have been developed and are used in different parts of the world: the Modified Mercalli scale (MM) is used in US, while the European Macroseismic Scale (EMS-98) in Europe, the Shindo scale in Japan, the MSK-64 scale in India, Israel, Russia and throughout the Commonwealth of Independent States, and the Liedu scale (GB/T 17742-1999) is used in mainland China; Hong Kong, on the other hand, uses the MM scale, and Taiwan uses the Shindo scale. Most of these scales have 12° of intensity, which are roughly equivalent to one another in values but may vary in the degree of sophistication employed in their formulation. Intensity ratings are classified into 12° of intensity in Roman numerals from I for insensible to XII for landscape reshaping at the high end. The Intensity Scale differs from the Richter Magnitude Scale in that the effects of any earthquake vary greatly from place to place, so there may be many Intensity values (e.g.: IV, VII) measured from one earthquake. Ratings of earthquake effects are based on the relatively subjective scale of descriptions (Chen et al., 1999).

Field observation shows that the seismic intensity of the meizoseismal area reaches up to XI for the Wenchuan earthquake (Fig. 5.1, the Liedu scale in this paper). The area is composed of two stripe-like zones along the earthquake surface rupture zone centering round Yingxiu Town of Wenchuan County and Qushan Town of Beichuan County, covering an area of 2,580 km². Among them, the Yingxiu intensity XI zone starts from Caijiagang Village near the Baihua bridge, Xuankou Town in the west, where landslide, debris flow and dammed lake were extremely developed, so that the site can be considered as the starting point of the macroscopic epicenter at the southwest end. The zone passes through the Yingxiu Town, north of the following places: Hongkou Township of Dujiangyan City, Xiaoyudong Town

H. Xing, X. Xu, *M8.0 Wenchuan Earthquake*, Lecture Notes in Earth Sciences 123, DOI 10.1007/978-3-642-01901-2_5, © Springer-Verlag Berlin Heidelberg 2011

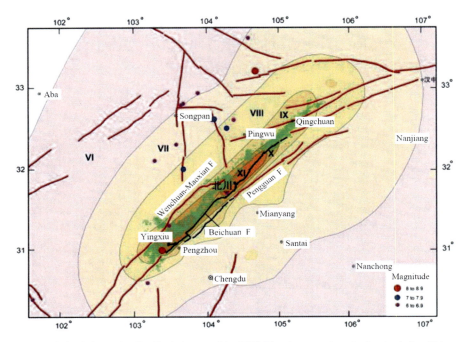

Fig. 5.1 Seismic intensity distribution map of the 2008 Wenchuan earthquake (revised after China Earthquake Administration, 2008). Legends: *Red solid line*: Active fault; *Black solid line*: Surface rupture zone associated with the Wenchuan earthquake; *Green circle*: Relocated aftershock

of Pengzhou City, Longmenshan Town, Hongbai Town of Shifang City and Jinhua Town, and then terminates at Qingping Township, Mianzhu City. The zone has a minor axis of about 17 km long and major axis of about 97 km long. The Beichuan intensity XI zone starts from Leigu Town, Beichuan County in the west and passes through Xuanping Township, Qushan Town, Chenjiaba Township, Pingtong Town of Pingwu County, south of Doukou Town and then terminates at the north of Nanba Town. It has a minor axis of 15 km and major axis of 71 km long (Fig. 5.1). The structures and buildings built within the two intensity XI zones were completely destroyed: the civil constructions in the zones were completely destroyed or seriously damaged, except that only a few structures or buildings were slightly damaged. The bridges in the zones were seriously damaged, from moderately damaged to totally collapsed. Tunnels remained mostly in good condition, and among the observed ten odd tunnels only two had been reinforced but can be passed through after the earthquake. Various kinds of high-tension towers and towers of communication base station at valley and flat land were in relatively good condition, while the towers on the slope were mostly damaged by collapsed rocks and landslide. Within the intensity XI zone several large industrial unit, such as the Xuankou Aluminum Plant, and some hydropower stations were seriously damaged. As the meizoseismal area is located in mountainous area, the phenomenon of liquefaction of sand as well as sand-blow and water eruption is scarcely observed.

The area of the intensity X zone is 3,650 km^2 and is slightly greater than that of the XI zone. The zone has a minor axis of about 23 km long and a major axis of about 224 km (including the intensity XI zone). Earthquake disasters in intensity X zone are extremely serious, but some relatively intact structures or buildings can still be observed there. The damage of the area in Hanwang Town, Mianzhu City, where the surface rupture zone passes through, is very serious; the buildings located within 1 km on both sides of the rupture zone, such as the Hanwang Infant School and the Middle School of the Dong Fang Steam Turbine Works were seriously destroyed. However, the earthquake disaster attenuates rapidly outside 1 km range. The intensity X zone of the Wenchuan earthquake has mostly reached into the piedmont alluvial and pluvial fans of the Longmenshan Mountains close to the margin of Sichuan Basin, but the phenomenon of sand-blow and water eruption is scarcely seen there.

The major axis of the intensity IX zone is 322 km long and the minor axis is 43 km (including intensity X and XI zones). The zone has an area of 8,730 km^2 (excluding the intensity X and XI zones). The zone coincides well with the surface rupture zone, and it just encloses almost the whole of coseismic rupture zone. Dujiangyan City and Qingchuan County town belong to typical intensity IX zone. In these region, ordinary residential houses (brick and concrete structured) were mostly seriously damaged, a part of brick and concrete structured houses were locally collapsed, a small part of multi-storey buildings were broken down and simply constructed houses were seriously collapsed, while the reinforced concrete structured houses constructed under newly issued building code were scarcely collapsed.

The zone of intensity VIII has a 430-km-long major axis, 106-km-long minor axis (including the zones of above intensity IX) and an area of 25,380 km^2, extending into Wenxian County, Wudu District and Kangxian County of Gansu Province, as well as the half part of Ningqiang County of Shaanxi Province. In this zone, ordinary residential houses were moderately or seriously damaged, simply-constructed houses were mostly collapsed, while the houses built according to seismic design were slightly or moderately damaged. The phenomena of sand blow and water eruption were relatively abundant in this zone. And especially on the foreland of the Shifang and Mianzhu Cities, ground caving in with an area of up to ten odd or several hundreds square meters can be observed at local places. Field observation indicates that the area of caving in is usually located on alluvial or pluvial fans. The bridges were slightly damaged, and the banisters were broken. Large-scale landslide and mudflow were rarely observed, but collapse and small-scale landslide were common, causing the damage of road in mountainous area.

5.2 Niu Juan Gou Hypocenter of Earthquake

From the field observation, Niu Juan Gou is regarded as the hypocenter or rupture initiation point of the Wenchuan earthquake. An indicator board (Fig. 5.2) has been set up in Niu Juan Gou (a gully) with the aerophotograph of terribly large scale

152	5 Earthquake Disasters and Damage Features of Structures and Buildings

Fig. 5.2 An indicator board of Hypocenter of the Wenchuan Earthquake set up after earthquake at Niu Juan Gou

landslide induced by the earthquake. Figure 5.3 is a brief introduction board on what happened around the hypocenter during the Wenchuan earthquake:

"At 14:28:04, May 12, 2008, an 8.0 Richter scale earthquake occurred about 11 km NW of Yingxiu Town. The focal depth was 14 km determined by seismic monitoring instruments as an accurate calculation of the mircoepicenter (latitude 31°, longitude 103°24′). This was located about 11 km NW of Yingxiu town, 7.5 km NW of Shuimo Town, and 1.7 km from Bajiao village of Xuankou town. The macro-epicenter was located at Niujuan beginning in the valley from Lianhua Xin to Caijia Gang Village of Xuankou town, an area of about 4 km^2".

When the earthquake occurred, the energy (the energy that 8.0 Ritcher scale earthquake was equivalent to 251 atomic bombs to explore simultaneously) erupted from one place called "Lianhua Xin". The sound was deafening and frightening with smoke rising from the rocky rolling stream. The rocks spurted from a waterfall what hit the opposite mountains and engendered a slope which was several hundred meters high. The rocks rebounded and undershot to the valley unceasingly. It formed a rock heap zone around 130 m wide, about 800 m thick. Because of rocks and sand obstructed an area 100 m wide, a 300-meter long cofferdam was formed on top of the accumulation zone. Millions of cubic meters of gravel buried more than 33 households claiming more than 23 people's lives". The related damages and earthquake lake can be clearly seen from the following photographs taken just after the earthquake (Figs. 5.4, 5.5 and 5.6).

5.3 Yingxiu Town

Fig. 5.3 Brief introduction of Niujuangou Hypocenter of the Wenchuan earthquake

5.3 Yingxiu Town

Yingxiu, was a beautiful town of Wenchuan County (Fig. 5.7), Sichuan Province. It is south of Wenchuan's county urban centre with an area of 115 km². It had a population of 6,906 before the earthquake. Yingxiu is just at the epicenter and one of the single worst hit areas of the May 12 Wenchuan earthquake (Seismic Intensity: XI). The ruins of Yingxiu Town was resulted from coseismic surface faulting, landslide and collapse along the seismogenic fault as shown in Figs. 5.8 and 5.9. The

Fig. 5.4 At Caijia Village, Baihua Township, Xuankou Town, a debris-flow began from the mouth of a gully at Baihua Bridge, running southwestward and then turning to nearly east-westward to reach Caijia Village, resulting in about 1-km-long mudflow deposits. Viewed to north

buildings on the earthquake surface rupture zone at Yingxiu Town were totally collapsed, which includes the existing concrete buildings (Fig. 5.10), the new buildings in construction (Fig. 5.11) and the school buildings (Figs. 5.12 and 5.13).

5.4 Qushan Town

Beichuan County, about 160 km northeast of Wenchuan, was at the center of one of two zones where seismic intensity was the highest at XI. The townships of Qushan and Leigu were hit particularly hard, with concrete structures crumbling to rubble under their own weight, or being crushed by landslides. About 80% of the buildings collapsed in the old town area and nearly 60% were levelled to the ground in the new town.

5.4 Qushan Town

Fig. 5.5 The NW-trending hanging gully at Xuankou Town converged with the Niujuan gully; the debris-flow come from landslide on the upper reach of the NW-trending gully. It rushed out of the mouth of the gully and reached the southern wall of the Niujuan Gully. The debris-flow deposits were as high as the mouth of the NW-trending gully. Viewed to north

Fig. 5.6 Debris-flow occurred between the mouth of a NW-trending hanging gully at Xuankou Town and Caijia Village had blocked the river channel to form a small dammed lake. Viewed to north

156 5 Earthquake Disasters and Damage Features of Structures and Buildings

Fig. 5.7 The beautiful Yingxiu Town before the Wenchuan earthquake

Fig. 5.8 The ruins of Yingxiu Town after the Wenchuan earthquake. The ruins resulted from coseismic surface faulting, landslide and collapse along the seismogenic fault

Qushan, the seat of Beichuan County, was also a beautiful town and surrounded by hills 500 m to 1,000 m high (Fig. 5.14). Qushan locates about 90 km from the epicenter of this earthquake and was one of the worst-hit towns by the earthquake. More than 8,600 of the 13,000 people living in the county seat were killed during the quake. Aerophotograph of the ruins of Qushan town after the Wenchuan earthquake shows that serious disasters were caused mainly by coseismic surface faulting along the seismogenic fault, large-scale landslides and collapses, as well as sand liquefaction at local sections (Figs. 5.15 and 5.16). Several typical damaged buildings were specially investigated. For examples, (a) the Wangjiayan landslide at the south of Qushan Town had buried 4 residential districts in Qushan Town, Beichuan County,

5.5 Road and Bridge Damages

Fig. 5.9 The ruins of Yingxiu Town after the Wenchuan earthquake. Viewed to north

as viewed at the different distances at the different locations (Figs. 5.17, 5.18 and 5.19). There were many public facilities in the path of Wangjiayan slide, including education bureau of Beichuan County, a public hospital, a kindergarten and a jail, which causing the death of 4,000 people; (b). A six-storey building in the north of Qushan Town, Beichuan County, remained only the first floor after the earthquake, while the above stories had been superimposed to make up a "sandwich structure" (Fig. 5.20); (c). The houses located on a normal fault scarp at Maoba Village, north of Qushan Town, Beichuan County were collapsed (the scarp on the right side is about 5 m high); two buildings on the hanging wall were tilted, and the tilting degree decreases with the increasing distance from the scarp (Fig. 5.21); (d) The collapsed Qushan Town was further destroyed/buried by the mud flow induced by the later storms (Fig. 5.17), which was clearly demonstrated through the comparison of the pictures taken before and after the earthquake (e.g. Figs. 5.22 and 5.23) and (e) the Beichuan middle school, where about 80% buildings collapsed, had two five-story teaching buildings collapsed (Fig. 5.24), causing the death of nearly half of its over 700 students.

5.5 Road and Bridge Damages

Besides of the road damage due to the surface ruptures and the landslides induced by the earthquake as described in Chap. 4, there exist some severe damages including: (a) various widely damaged bridges (e.g. Figs. 5.25, 5.26, 5.27, 5.28, 5.29 and 5.30); (b) severely deformed railway (Fig. 5.31) and (c) road damaged/blocked by the huge rock fall due to the earthquake (Figs. 5.32, 5.33 and 5.34).

158　　5 Earthquake Disasters and Damage Features of Structures and Buildings

Fig. 5.10 The buildings on the earthquake surface rupture zone at Yingxiu Town were totally collapsed

5.5 Road and Bridge Damages

Fig. 5.11 The second storey of a six-storey building in construction on the footwall was lost after the earthquake. Viewed to west

Fig. 5.12 The main teaching building of Yingxiu Middle School at Yingxiu Town, Wenchuan County, was collapsed by the earthquake. Viewed to north

Fig. 5.13 The dormitory building of the Yingxiu middle school at Yingxiu Town, Wenchuan County, was damaged by X-shaped shearing cracks, and the first floor was lost after the earthquake. Viewed to north

5.5 Road and Bridge Damages

Fig. 5.14 The beautiful Qushan town before the Wenchuan earthquake

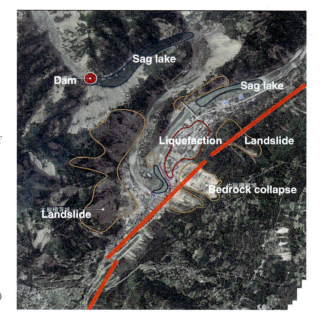

Fig. 5.15 Aerophotograph of the ruins of Qushan Town after the Wenchuan earthquake. Serious disasters were caused mainly by coseismic surface faulting along seismogenic fault, large-scale landslides and collapses, as well as sand liquefaction at local sections (by courtesy of Institute of remote sensing applications, Chinese academy of sciences)

Fig. 5.16 The collapsed Qushan Town by the earthquake (taken before the storm)

Fig. 5.17 The collapsed Qushan Town by the earthquake and storm induced muddy flow

5.6 Other Damages

Besides of those damages mentioned above, some special geological phenomena were also observed, such as earthquake-induced pitfall (Fig. 5.35), sand-blow and water eruption (Fig. 5.36), mole-tracks (Fig. 5.37) and earthquake lakes (e.g. Figs. 5.38 and 5.39). The most famous earthquake lake, Tangjiashan Quake Lake, was caused by a landslide on Mount Tangjiashan which dammed the Jianjiang River and created the Tangjiashan Quake Lake. The lake was once in danger of causing the Tangjiashan Dam to collapse and catastrophically flood downstream communities, totaling over a million persons. On June 10, 2008, the lake spilled through an artificially constructed sluice channel and flooded the evacuated town. No casualties were caused due to Wenchuan earthquake induced lakes.

5.6 Other Damages

Fig. 5.18 Wangjiayan landslide south of Qushan Town had buried 4 residential districts in Qushan Town, Beichuan County, causing the death of more than 4,000 people. Viewed to southwest

Fig. 5.19 The impact wave of the front of Wangjiayan landslide had caused the collapse of houses into ruins (close view) at Qushan Town, Beichuan County. The collapse of mountain had also caused the collapse of newly built Beichuan Middle School and the adjacent houses (in distance). Viewed to north

Fig. 5.20 A six-storey building in the north of Qushan Town, Beichuan County, remained only the first floor after the earthquake, while the above stories had been superimposed to make up a "sandwich structure". Viewed to east

Fig. 5.21 The houses located on a normal fault scarp at Maoba Village, north of Qushan Twon, Beichuan County, were collapsed (the scarp on the right side is about 5 m high); two buildings on the hanging wall were tilted, and the tilting degree decreases with the increasing distance from the scarp. Viewed to southwest

5.6 Other Damages

Fig. 5.22 Photographs of the outdoor gate of the Beichuan government taken at different time: (**a**) before the earthquake; (**b**) just after the earthquake but before the heavy storm in June 2008 and (**c**) after the storm, the gate was almost buried by the muddy flow

Fig. 5.23 The concrete buildings far from the fault zone in the center of Qushan town were survived during the Wenchuan earthquake (**a**), but buried by the later storm induced muddy flow (**b**)

Fig. 5.24 The collapsed teaching buildings at Beichuan middle school

Fig. 5.25 The middle part of a bridge near Gaoyuan Village, Hongkou Township, Dujiangyan City was broken down. Viewed to northwest by north

5.7 What Learned from Wenchuan Earthquake

Fig. 5.26 Slope collapse and fracturing of railway bridge at Muguaping Phosphorus Mine, Pengzhou City. Viewed to north

5.7 What Learned from Wenchuan Earthquake

Since the seismogenic fault of the Wenchuan earthquake is the Beichuan fault and the Pengguan fault of the Longmenshan thrust belt, which are thrust faults with right-slip component, the surface ruptures or strong deformation display significant difference on the hanging-wall and footwall of the thrust fault. As a result, the hanging-wall effect of earthquake disaster is significant: the deformation and subsidiary fractures or fissures were less developed on the footwall, and only secondary reverse faults or mole-tracks were occasionally observed; numerous open fractures or cracks were widely developed on the hanging wall near the fault, and the areas of high intensity are mostly located on the hanging wall of the surface rupture zone. Field measurements show that the width of the surface rupture zones, including the main ruptures and secondary tension cracks, associated with the Wenchuan earthquake is relatively narrow, normally within the range of 21–45 m, and at individual site such as the yard of the Bailu Central School it is only 13 m (Fig. 5.40). It should be pointed out, however, that the ground, the two-storey building of Mr. Dong's family built in 2006 on the hanging wall of ~4 m-high pressure ridge just 36 m away from the main scarp at Shenxigou, Hongkou Village west of the Beichuan surface rupture zone, were preserved pretty well during the earthquake, and no obvious cracks occurred on the ground, walls and floor-slabs of the building. Similar phenomena can also be observed on the Hanwang rupture zone at Laojie Village, Bailu Town. For example, the one-storey house of Mr. Zhu's family built in 2005 on the hanging wall of the fault just 35 m away from the scarp was slightly fractured on

168　　5 Earthquake Disasters and Damage Features of Structures and Buildings

Fig. 5.27 The Xiaoyudong surface rupture zone obliquely cuts the river valley of the Baishuihe River and the eastern seat of the Xiaoyudong Bridge. The bridge on the hanging wall was displaced into several sections. Viewed to west

its wall during the earthquake. The aforementioned phenomena of surface ruptures and building damages may indicate that the Wenchuan earthquake surface rupture zone is characterized by deformation localization. Correspondingly, the width of the damage zone of surface constructions and buildings caused by co-seismic faulting along the seismogenic fault is constrained within the width range of the earthquake surface rupture zone (e.g. Figs. 5.41, 5.42, 5.43 and 5.44). Almost all of the surface constructions built across fault scarps were completely collapsed or structurally damaged by the shock, while those not across the fault scarps such as the teaching building at Bailu School was not structurally damaged (Fig. 5.45). In-situ measured width of the damage zone of surface constructions near the fault scarp varies from 13 to 36 m, indicating that the co-seismic faulting along the earthquake surface rupture zone may overrun all fortifications.

5.7 What Learned from Wenchuan Earthquake

Fig. 5.28 The earthquake site of the Sino-French Bridge to the west of Bailu School in Bailu town

Fig. 5.29 Beichuan Bridge plate was shifted northeastward, whereas the bridge pillars were not displaced (view to NW)

Fig. 5.30 The entire length of Dujiangyan-Yingxiu-Wenchuan express way along the bank of the Minjiang River was broken. Viewed to north

Fig. 5.31 Flexural deformation of the rails of the Muguaping Phosphorus Mine railway, Shifang City. Viewed to north

5.7 What Learned from Wenchuan Earthquake 171

Fig. 5.32 An isolated giant fallen rock and a sag on its side at Yingxiu Town, Wenchuan County. Viewed to northeast

Fig. 5.33 The fallen rocks damaged the vehicles and blocked the Gaochuan road in Anxian County

Fig. 5.34 The pavement of a road at Shenxigou Village, Hongkou Township, Dujiangyan City was covered with a large number of big fallen rocks. Viewed to north

Fig. 5.35 An earthquake-produced pitfall of about 3 m in diameter at Fangshi Village, Maluba township, Guangyuan City. Viewed to south

5.7 What Learned from Wenchuan Earthquake

Fig. 5.36 Remains of sand-blow and water eruption at Shehong Village. Viewed to west

Fig. 5.37 Mole-track developed at Shenxigou Village, Hongkou Township, Dujiangyan City

Fig. 5.38 An earthquake lake in the north of Hongbai Town, Shifang City. Viewed to west

Fig. 5.39 The dam of a dammed lake at a site between the Suishui Village, Anxian County and Gachuan Town. Viewed to southwest

5.7 What Learned from Wenchuan Earthquake

Fig. 5.40 Structures and width of secondary tension-crack associated with the Wenchuan earthquake surface ruptures and damage features of surface constructions and buildings: (**a**) Sub-parallel tension-cracks developed on top of a simple pressure ridge on the Beichuan surface rupture zone at Shenxigou Village, Hongkou Township, Dujiangyan City; viewed to southwest; (**b**) Tension-crack parallel to the strike of the main scarp developed on a back-thrust pressure ridge on the Beichuan-Yingxiu surface rupture zone at Muerdi Village on the upper floodplain of the Pingtong River, Pingtong Town; the main scarp is about 4.2 m high; the trail was right-laterally offset 3.7 m, and the width of surface rupture zone is 36 m; viewed to northwest; (**c**) Tension-cracks densely concentrated within 45 m width on top of an about 3.4-m-high simple thrust fault scarp on the western end of the NW-trending Xiayudong surface rupture zone to the north of Luoyuan Village; viewed to south; (**d**) East-west-trending en-echelon tension-cracks developed on the uplifted side of the scarp at a site where ~0.9-m-high pressure ridge on the Beichuan-Yingxiu surface rupture zone cuts across the road at Lizikan VillageNanba Town, Pingwu County; the surface rupture zone is 29 m wide; viewed to northeast by north; (**e**) The width of a fault-related fold scarp and damages of buildings across the scarp on the Hanwang-Bailu surface rupture zone at Wangjiaba Village; the scarp is N50°E-trending, and the surface rupture is 21 m wide; viewed to northwest; (**f**) Simple pressure ridge on the Beichuan-Yingxiu surface rupture zone in the north of Yingxiu Town and the damages of cement dikes on the western bank of the Minjiang River; the scarp is N70°E-trending and the surface rupture zone is 34 m wide; viewed to

Fig. 5.40 (continued) northwest; (**g**) Simple pressure ridge on the Hanwang-Bailu surface rupture zone at Bailu Middle School, Pengxian County; the teaching building on the footwall of the scarp was fractured but not collapsed, and the building on the hanging wall was relatively intact; the scarp is N50°E-trending, and the width of the surface rupture zone is 13 m; viewed to northeast; and (**h**) The width of the simple thrust fault scarp and the collapsed buildings on the NE-ternding Xiaoyudong surface rupture zone on the south street in Xiaoyudong Town; the width of the surface rupture zone is 36 m; viewed to northwest

Fig. 5.41 The houses located on a normal fault scarp at Maoba Village, north of Qushan Twon, Beichuan County, were collapsed (the scarp on the right side is about 5 m high). Two buildings on the hanging wall were tilted, and the tilting degree decreases with the increasing distance from the scarp. Viewed to southwest

5.7 What Learned from Wenchuan Earthquake

Fig. 5.42 A storied building located across an earthquake scarp (the building in distance) at Lizikan Village, Pingwu County, was structurally damaged and tilted, and the storied building on the hanging wall was tilted. Viewed to north

Fig. 5.43 A wood framework structured house at Shenxigou Village, Hongkou Township, Dujiangyan City, was tilted with the declination of fault scarp, and the upper wall was slightly tilted with the tilting of the ground. Viewed to west

Fig. 5.44 The Bajiaomiao Sanatorium at Hongkou Twonship, Dujiangyan City, located on the fault scarp was collapsed, but the part on the hanging wall was fractured and relatively intact. Viewed to northeast

Fig. 5.45 The teaching building of the Bailu School in Pengxian County on the hanging wall of the scarp was relatively intact after the earthquake (**a**) with comparison of that before earthquake (**b**)

5.7 What Learned from Wenchuan Earthquake

The damages of buildings in the other sections were caused either by the shock of ground, the bad geologic conditions of their position (located at landslide, mud- or debris-flow and collapse area, etc), or by the ignorance of seismic design for constructions and the bad quality of construction.

A large number of seismic events show that the active fault is not only the generating source of earthquake, but also the main cause of seismic disaster. In General, an earthquake of above magnitude 7 may produce about several meters coseismic surface displacement, while the current seismic design is still unable to prevent the direct damage of surface constructions and buildings caused by such a large surface displacement. Serious earthquake disaster is usually distributed along a narrow zone on both sides of the seismogenic fault (Xu et al., 1996). For examples, the areas of serious disaster caused by the M7.2 Hanshin-Awaji earthquake, Japan in 1995 were concentrated along the strand of seismic fault, the Nojima-Suwayama-Suma fault. According to the report of the Ashahi Sciences Japan, more than 90% of deaths and the area, where more than 30% of wood-structured houses were collapsed, are distributed mainly within 3-km-wide range on both sides of the seismic fault. The maximum peak acceleration as recorded by the Kobe Marine Weather Observatory near the epicenter reached up to 0.818 g. The zones of serious disaster produced by the M 7.8 Izmit earthquake, Turkey in 1999 were concentrated along the northern branch fault of the western segment of the North Anatolian fault. The constructions and buildings located on the seismic fault were totally collapsed, while the main structures of constructions located several tens meters away from the fault on both sides were only slightly damaged. Along the seismogenic fault (the Chelongpu fault) of the M 7.6 Chichi earthquake, Taiwan area, the surface ruptures had destroyed all buildings located on the fault or ten odd meters away from the fault on both sides, while the buildings located outside this range preserved to be relatively intact (Li et al., 2000). The maximum peak acceleration near the seismic fault reached up to 1 g, and attenuated rapidly toward the both sides of the fault (Wang et al., 2001). Obviously, the seismic design cannot simply used to mitigate the damages of constructions and utility lines projects on active fault or earthquake surface rupture zones. The first countermeasure should be taken for disaster mitigation is to avoid the displacement zone of active fault, while the constructions outside the displacement zone may greatly decrease the damages caused by the effects of earthquake shocking by adopting higher seismic design. Therefore, the delineation of the correct location of active fault with earthquake potential, so that the major projects, utility lines constructions and residential buildings can avoid the active fault with earthquake potential, may decrease effectively the losses caused by the future earthquake (Xu et al., 2002).

The government of the California State, which is an earthquake-prone region, has paid great attention to the disaster mitigation projects on active faults. Not only the strict seismic design for building construction has been established, but also the laws and regulations for disaster mitigation related to active faults have been issued to normalize the earlier-stage appraisals of active faults for engineering project, so that the newly-built constructions may avoid the active faults and to prevent the direct damages caused by the possible coseismic surface displacement along the active fault. As early as in 1971, when the San Fernando earthquake occurred, the scientists and officials from California State have noticed the direct earthquake disasters caused by the active fault, and a "Special Studies Zones" was adopted by the state government at the beginning of 1972, which was revised into the "earthquake Fault Zoning Act" after the occurrence of the Northridge earthquake in 1994. The main purpose of these regulations is to prohibit the construction of residential houses on the surface appearance of the active fault. The main contents of this regulation are: the state government announces the zoning of active fault according to a certain procedures; if one development project involves this zoning of active fault, then professional technicians should be entrusted to carry out detailed geologic investigation and appraisal of the active fault, and to composed a detailed investigation report; if active fault is found, then the residential houses should be built more than 15 m away from the fault on both sides, but a single wood-structured or steel-framework house not exceeding to two-storey is not constrained by this regulation. In addition to the "Earthquake Fault Zoning Act", a series of public welfares have been promoted as the disaster mitigation action of the country or local governments. For examples, the 1: 24,000 mapping project of active fault in California region has been carried out as a basic countermeasure for mitigating the disaster related to active faults, and now the mapping of 547 outcropped active faults have been fulfilled.

In Taiwan area of China, the article 5, Sect. 2 of the management measures for constructions and development in slope area stipulates that: the site where the geologic structures are bad, the strata are broken, the active fault or the consequent slope might be slipped, should not be constructed and exploited. After the Chichi earthquake of 1999, facing the problem of whether to reconstruct the buildings destroyed directly by the fault, the Construction Office of Taiwan's authorities has adopted the basic principle of prohibiting and restricting permanently the reconstruction of these buildings: in the area with distinct active fault strand, within 15 m wide range on both sides of the fault strand and in a part of areas where the active fault is still unclear, the construction of public utilities and major public business arena, such as school, hospital, police station, fire control and disaster rescue organs are prohibited, while on the private land only one- or two-storey buildings are allowed to be constructed.

The Wenchuan earthquake of May 12, 2008 that struck the Longmenshan region on the middle segment of the North-South seismic belt of China has caused tremendous disasters. At the same time, the earthquake has revealed the following distribution features of disasters produced by steeply-dipping right-lateral strike-slip thrust fault: the width of the Wenchuan earthquake surface rupture zone is 45 m, and may reach up to 100 m or more at individual sections; the surface rupture zone is characterized by deformation localization (Xu et al., 1996, 2002, 2008a and 2008b); the meaizoseismal area of the Wenchuan earthquake is located in Longmenshan mountainous areas, including Yingxiu, Beichuan and Wenchuan town and county; High seismic intensity regions involve the Dujiangyan, Mianzhu, Shifang, Qingchuan and Pengzhou Cities. Among them, Beichuan and Yingxiu suffered from devastating disasters, where the rate of damaged and destroyed buildings reached up to 90% or more and the casualties are very heavy. Dujiangyan, Mianzhu

5.7 What Learned from Wenchuan Earthquake

Cities and Hanwang Town suffer from very serious damages. The following reasons have caused so serious disaster and heavy loses (Deng, 2008):

(1) The Wenchuan earthquake is characterized by great magnitude, high energy shallow focal depth and long duration time of rupturing (about 90 s) (Chen Y et al., 2008b; Ji et al 2008; Nishimura and Yagi 2008; Yagi 2008); ground motion in meizoseismal area was very vigorous, and the maximum peak ground acceleration recorded by the Wolong Strong Motion Observation Station reached up to 956.7 Gal (Li et al., 2008c);

(2) The seismogenic fault outcrops onto the surface and passes directly through these cities and towns to form 240-km-long and several-ten-meter-wide surface rupture zone or fold deformation zone with surface displacement of 6.5 ± 0.5 m; the buildings on the earthquake surface rupture zones were completely destroyed into ruins, while the intensive shock near the fault had enhanced the damages of these cities and towns;

(3) The cities and towns in the mountainous areas are located mostly at a long and narrow valley, along which active fault is developed. River or stream runs along the valley, while the city or town is usually located at a bad site such as the lower terrace or floodplain of the river, on both sides of which there are cliff and steep slope with slope angle of up to 50–60°, or even 70–80°. During the earthquake these unstable relief had caused large scale collapses and landslides, which buried the cities and towns, dammed the rivers and blocked the roads. The front of landslide produced strong impact wave destroying all the buildings in front of the landslide to form a circular damage zone, which caused the enlargement of calamities and extremely heavy casualties. Wangjiayan landslide in the south and large-scale collapses in the north of Qushan Town, Beichuan County, as well as the NE-trending seismic fault passing through the town are the main reasons that cause the complete ruin of the town;

(4) The other secondary hazards had also strengthened the earthquake disaster. For examples, a rainstorm may destroy the loose hills and vegetation that had been damaged by the earthquake, resulting in mudflow that may cause tremendous secondary calamities. The rainstorm occurred on September 24, 2008 in the Wenchuan earthquake area had produced a large-scale mudflow that buried the Beichuan old county town;

(5) The standard of seismic design in the earthquake area was too low. Before the earthquake, the seismic design was fortified against seismic intensity VII and ground acceleration of 0.1 g, while the houses in the rural areas were not fortified. Therefore, when the shock came, the buildings were lacking defending capability, so that they were damaged or destroyed;

The following inspirations for earthquake disaster mitigation can be drawn from the detailed field investigations and measurements of the Wenchuan earthquake:

(1) The appraisals of construction environment, as well as seismic and geologic safety assessments should be carried out before the constructions of cities and

towns. Even though the construction environments of constructed city or town should also be reexamined and reappraised, to avoid certainly the distribution zones of active fault and active tectonics. In mountainous areas, the cities and towns should not be constructed at a steeply-dipping landform or relief and at a narrow valley; great attention should be paid to the existence and distribution of new and old collapse or landslide; the construction site should avoid the potential landslides, while the existing landslide should be tackled; cares should be taken to avoid the harmful high-angle dissected slope in the construction of city or town. In plain and piedmont areas, the bad construction sites, such as the site where sand liquefaction is easily to occur and the site where underground karst cave is developed should be avoid. Propaganda and popularization should be carried out in the rural areas to promote the selection of safe construction environment even for discrete housing.

(2) The construction of city and town should follow strictly the standards of seismic design issued by the country. Although precise seismic zoning is still an unresolved problem, and the magnitude of actual earthquake may exceed the seismic risk expected by the seismic zoning map, a large number of actual cases of earthquake disasters both at home and abroad, including the case of the Wenchuan earthquake, indicate that fortifying the constructions according to the standards of seismic design for the locality may greatly mitigate earthquake disaster.

(3) In the construction, technical specifications of the relevant Building Code must be implemented. Reasonable technical measures should be adopted to strengthen the structural links between the components and to enhance the entirety of the constructions; the corresponding measures should be adopted to raise the anti-seismic loading and anti-deformation capacity of the walls of the constructions.

Ancient Chinese proverb said: the overturned cart ahead is a warning to the cart behind. At present, no active fault act has been issued by the government, except that for major project related to the nuclear power station and nuclear waste disposal sites. It is proposed, therefore, that the relevant departments of the government carry out the legislation and management of active fault act; large scale mapping of active faults in China's mainland should be progressively carry out, in order to provide scientific basis for avoiding the active faults in land planning and utilization, as well as in constructions; mitigate passively, initiatively and effectively the potential earthquake disasters caused by active faults.

References

Allen C. R., Luo Z., Qian H., Wen X., Zhou H. Huang W., 1991. Field study of a highly active fault zone: the Xianshuihe fault of southwestern China, Geol. Soc. Am. Bull., 103: 1178–1199.

Avouac J. P., Tapponnier P., 1993. Kinematic model of active deformation in Central Asia, Geophys. Res. Lett., 20: 895.

Basile C., Brun J. P., 1999. Transtensional faulting patterns ranging from pull-apart basins to transform continental margins: an experimental investigation, J. Struct. Geol., 21(1): 23–37.

Burchfiel B. C., Royden L. H., van der Hilst R. D., Chen Z., King R. W., Li C., Lu J., Yao H., Kirby E., 2008. A geological and geophysical context for the Wenchuan earthquake of 12 May 2008, Sichuan, People's Republic of China, GSA Today, 18(7), doi:10.1130/GSATG18A.1.

Chen S. F., Wilson C. J. L., Deng Q. D., Zhao X. L., Luo Z. L., 1994. Active faulting and block movement associated with large earthquakes in the Min Shan and Longmen mountains, northeastern Tibetan plateau, J. Geophys. Res., 99: 24,025–24,038.

Chen D., Shi Z., Xu Z., Gao G., Nian J., Xiao C., Feng Y., 1999. China seismic intensity scale (in chinese). General administration of quality supervision, inspection, and quarantine of P.R.C.. http://www.dccdnc.ac.cn/html/zcfg/gfxwj2.jsp. Retrieved 2008-09-12.

Chen G., Ji F., Zhou R. et al., 2007. Primary research of the activity segmentation of Longmen Shan fault zone since late quaternary, Seismol. Geol., 29(3): 657–673 (in Chinese).

Chen J., Shao G., Lu Z., Hudnut K., Jiu J., Hayes G., Zeng Y., 2008a. Rupture history of 2008 may 12 Mw8.0 Wenchuan earthquake: Evidence of slip interaction, EOS Trans. AGU, 89(53), Fall Meet. Suppl., Abstract S23E–02.

Chen Y. T., Xu L. S., Zhang Y., Du H. L., 2008b. Report on great Wenchuan earthquake source of May 12, 2008, see http://www.csi.ac.cn/sichuan/chenyuntai.pdf (in Chinese).

Chen J. H., Liu Q. Y., Li S., Guo B., Li Y., Wang J., Qi Sh., 2009a. Seismotectonic study by relocation of the Wenchuan Ms8.0 earthquake sequence, Chin. J. Geophy., 52(2): 390–397.

Chen G. H., Xu X. W., Yu G. H., An Y. F., Yuan R. M., Guo T. T., Gao X., Yang H., Tan X. B., 2009b. Co-seismic slip and slip partitioning of multi-faults during the Ms8.0 2008 Wenchuan earthquake, Chin. J. Geophy., 52(5): 1384–1394.

China Earthquake Administration, 1999. The Catalogue of Modern Earthquakes in China. China Science and Technology Press, Beijing, p. 637.

China Earthquake Administration, 2008. http://www.cea.gov.cn/manage/html/8a8587881632fa5c 0116674a018300cf/_content/08_08/29/1219979564089.html.

Deng Q. (ed.), 2007. Active Tectonics Map of China. Seismological Press, Beijing (in Chinese).

Deng Q., 2008. Some thoughts on the Ms8.0 Wenchuan, Sichuan earthquake, Seismol. Geol., 30(4): 811–827.

Deng Q., Chen S., Zhao X. L., 1994. Tectonics, seismicity and dynamics of Longmenshan mountains and its adjacent regions, Seismol. Geol., 16(4): 389–403 (in Chinese).

Densmore A. L., Ellis M. A., Li Y., Zhou R., Hancock G. S., Richardson N., 2007. Active tectonics of the Beichuan and Pengguan faults at the eastern margin of the Tibetan Plateau. Tectonics, 26, TC4005, doi:10.1029/2006TC001987.

Dieterich J. H., 1978. Time-dependent friction and the mechanics of stick-slip, Pure Appl. Geophys., 16: 790–806.

Dieterich J. H., 1979. Modeling of rock friction 1. Experimental results and constitutive equations, J. Geophys. Res., 84: 2161–2168.

Harris R. A., Day S. M., 1999. Dynamic 3D simulations of earthquakes on en echelon faults, Geophy. Res. Lett., GL900377, 26(14): 2089–2092.

He H., Oguchi T., 2008. Late quaternary activity of the Zemuhe and Xiaojiang faults in Southwest China from geomorphological mapping, Geomorphology, 96: 62–85.

Huang Y., Zhang T., 2008. Relocation of M8.0 Wenchuan earthquake and its aftershock sequence, Sci. China Ser. D Earth Sci., 51: 1703–1711.

Ji C., Hayes G.. Preliminary result of the May 12, 2008 Mw 7.9 eastern Sichuan, China earthquake, U.S. Geol. Surv., Va. (available at http://earthquake.usgs.gov/eqcenter/eqinthenews/2008/us2008ryan/finite_fault.php)

Jones L. M., Han W., Hauksson E., Jin A., Zhang Y., Luo Z., 1984. Focal mechanisms and aftershock locations of the Songpan earthquakes of August 1976 in Sichuan, China, J. Geophys. Res., 89(B9): 7697–7707, doi:10.1029/JB089iB09p07697.

King R. W., Shen F., Burchfiel B. C., Royden L. H., Wang E., Chen Z., Liu Y., Zhang X. -Y., Zhao J. -X., Li Y., 1997. Geodetic measurement of crustal motion in Southwest China, Geology 25: 179–182.

King G., Klinger Y., Bowman D., Tapponnier P., 2005. Slip-partitioned surface breaks for the Mw7.8 2001 Kokoxili earthquake, China, Bull. Seismol. Soc. Am., 95: 731–738.

Lettis W., Bachhuber J., Witter R. et al., 2002. Influence of releasing step-overs on surface fault rupture and fault segmentation: examples from the 17 august 1999 Izmit earthquake on the North Anatolian fault, Turkey, Bull. Seismol. Soc. Am., 92(1): 19–42.

Li X., Zheng J., Liao Q., 2000. Site deformation of the Jiji earthquake of Sep. 12 and the problem of prohibition and restriction of building construction on active fault. In: Proceedings of the 8th Symposium on the Geophysics in Taiwan area. pp. 669–675.

Li Y., Zhou R., Densmore A. L., Elli M. A., 2006. Geomorphic evidence for the late Cenozoic strike-slipping and thrusting in Longmen mountain at the eastern margin of the Tibetan plateau, Quat. Res., 26(1): 40–51.

Li H., Fu X., Van der Woerd J., Si J., Wang Z., Hou L., Qiu Z., Li N., Wu F., Xu Zh., Tapponnier P., 2008a. Co-seisimic surface rupture and dextral-slip oblique thrusting of the Ms 8.0 Wenchuan earthquake, Acta. Geol. Sin., 82(12): 1623–1643.

Li Y., Zhou R., Densmore A., Yan L., Richardson N., Dong S., Ellis M. A., Zhang Y., He Y., Chen H., Qiao B., Ma B., 2008b. Surface rupture and deformation of the Yingxiu-Beichuan fault by the Wenchuan earthquake, Acta. Geol. Sin., 82(12): 1688–1702 (in Chinese).

Li X., Zhou Zh., Yu H., Wen R., Lu D., Huang M., Zhou Y., Cu J., 2008c. Strong motion observations and recordings from the great Wenchuan earthquake, Earthquake Eng. Eng. Vib., 7(3): 235–246 doi:10.1007/s11803–008-0892–x.

Lin M. B., Wu S., 1991. Deformation characteristic of the napper tectonic belt in Longmen Mountains, J Chengdu Inst. Technol., 18(1): 46–55.

Lin A., Ren Zh., Jia D., Wu X., 2009. Co-seismic thrusting rupture and slip distribution produced by the 2008 Mw 7.9 Wenchuan earthquake, China, Tectonophysics, doi:10.1016/j.tecto.2009.02.014.

Liu Q. Y., Chen J., Li S., Li Y., Guo B., Wang J., Qi S., 2008. The Ms8.0 Wenchuan earthquake: preliminary results from the Western Sichuan mobile seismic array observations, Seismol. Geol., 30: 584–595.

Liu Q., Li Y., Chen J., Guo B., Li S., Wang J., Zhang X., Qi S., 2009. Wenchuan MS8.0 earthquake: preliminary study of the S-wave velocity structure of the crust and upper mantle, Chin. J. Geophys., 52(2): 309–319.

Lu J. N., Shen Z., Wang M., 2003. Velocity filed and tectonic block division of crustal movement obtained by GPS measurements in Sichuan Yunnan region, Seismol. Geol., 25(4):543–554 (in Chinese).

References

Ma B., Su G., Hou Z. et al., 2005. Late quaternry slip rate in the central part of the Longmen Shan fault zone from terrace deformation along the Minjiang river[J], Seismol. Geol., 27(2): 234–242 (in Chinese).

Meade B., 2007. Present-day kinematics at the India-Asia collision zone, Geology, 35: 81–84, doi: 10.1130/G22924A.1.

Nishimura N., Yagi Y., 2008. Rupture process for May 12, 2008 Sichuan earthquake: (preliminary result): http://www.geol.tsukuba.ac.jp/~nisimura/20080512/ (June 2008).

Parsons T., Ji C., Kirby E., 2008. Stress changes from the 2008 Wenchuan earthquake and increased hazard in the Sichuan basin, Nature, 454: 509–510, doi:10.1038/nature07177.

Ruina A. L., 1983. Slip instability and state variable friction laws, J. Geophys. Res., 88: 10359–10370.

Shen Z., Lu J., Wang M., Burgmann R., 2005. Contemporary crustal deformation around the southeast borderland of the Tibetan plateau, J. Geophys. Res., 110: 1–17.

Song H., 1994. The comprehensive interpretation of geological and geophysical data in the orogenic belt of Longmen mountains, China, J. Chengdu Inst. Technol., 21(2): 79–88 (in Chinese).

Song F., Wang Y., Yu W., Cao Z., Sheng X., Shen J., 1998. Xiaojiang Fault Zone, Seismological Press, Beijing (in Chinese).

Tamer Y. D., Omer E., Ahmet D. et al., 2005. Step-over and bend structures along the 1999 Duzce earthquake surface rupture, North Anatolian fault, Turkey, Bull. Seismol. Soc. Am., 95(4): 1250–1262 doi:10.1785/0120040082.

Tang R. Han W. (eds), 1993. Active Faults and Earthquakes in Sichuan Province. Seismological Press, Beijing, p. 368 (in Chinese).

Tapponnier P., Molnar P., 1977. Active faulting and tectonics in China, J. Geophys. Res., 82: 2905–2930.

Tapponnier P., Peltzer G., Le Dain A. Y., Armijo R., Cobbold P., 1982, Propagating extrusion tectonics in Asia: new insights from simple experiments with plasticine, Geology, 10: 611–616.

Tapponnier P., Xu Zh., Roger F., Meyer B., Arnaud N., Wittlinger G., Yang J., 2001. Oblique stepwise rise and growth of the Tibet plateau, Science, 294: 1671–1677.

Tim D., Ken Mc., 1997. Analog modeling of pull-apart basins, AAPG Bull., 81(11): 1804–1826.

Toda S., Lin J., Meghraoui M., Stein R. S., 2008. 12 may 2008 M = 7.9 Wenchuan, China, earthquake calculated to increase failure stress and seismicity rate on three major fault systems, Geophys. Res. Lett., 35(L17305), doi:10.1029/2008GL034903.

Wang W. H., Chang Sh. H., Chen Ch. H., 2001. Fault slip inverted from surface displacements during the 1999 Chi-Chi, Taiwan, earthquake, Bull. Seismol. Soc. Am., 91(5): 1167–1181.

Wang X. Y., Zhu W. Y., Fu Y., 2002. Present time crustal deformation in China and its surrounding regions by GPS, Chin. J. Geophys., 45(2): 198–209 (in Chinese).

Wang W. M., Zhao L. F., Li J., Yao Zh. X., 2008. Rupture process of the Ms 8.0 Wenchuan earthquake of Sichuan, China, Chin. J. Geophy., 51(5): 1403–1410.

Wang Q., Cui D., Zhang X., Wang W., Liu J., Tian K., Song Zh., 2009. Coseismic vertical deformation of the MS8.0 Wenchuan earthquake from repeated levelings and its constraint on listric fault geometry, Earthquake Sci., 22: 595–602, doi:10.1007/s11589-009-0595-z.

Wen X., 1993. Rupture segmentation and assessment of probabilities of seismic potential on the Xiaojiang fault zone, Acta. Seismol. Sin., 6: 993–1004.

Wen X., 2000. Character of rupture segmentation of the Xianshuihe-Anninghe-Zemuhe fault zone, western Sichuan, Seismol. Geol., 22: 239–249 (in Chinese).

Wen X., Ma S., Xu X., He Y., 2008. Historical pattern and behavior of earthquake ruptures along the eastern boundary of the Sichuan-Yunnan faulted-block, southwestern China, Phys. Earth Planet Inter., 168: 16–36.

Xing H. L., Makinouchi A., 2002a. Finite element modeling of multibody contact and its application to active faults, concurrency and computation, Pract. Exp., 14: 431–450.

Xing H. L., Makinouchi A., 2002b. Three dimensional finite element modelling of thermomechanical frictional contact between finite deformation bodies using R-minimum strategy, Comput. Methods Appl. Mech. Eng., 191: 4193–4214.

Xing H. L., Makinouchi A., 2002c. Finite element analysis of sandwich friction experimental model of rocks, Pure Appl. Geophys., 159: 1985–2009.

Xing H. L., Mora P., 2006. Construction of an intraplate fault system model of South Australia, and simulation tool for the iSERVO institute seed project, Pure Appl. Geophys., 163: 2297–2316.

Xing H. L., Zhang J., 2009. Finite element modeling of non-linear deformation behaviours of rate-dependent materials using an R-minimum strategy, Acta Geotech., 4: 139–148, doi:10.1007/s11440-009-0090-7.

Xing H. L., Mora P., Makinouchi A., 2004. Finite element analysis of fault bend influence on stick-slip instability along an intra-plate fault, Pure Appl. Geophys., 161: 2091–2102.

Xing H. L., Mora P., Makinouchi A., 2006. An unified friction description and its application to simulation of frictional instability using finite element method, Philos. Mag., 86: 3453–3475.

Xing H. L., Makinouchi A., Mora P., 2007a. Finite element modeling of interacting fault system, Phys. Earth Planet. Inter., 163: 106–121, doi:10.1016/j.pepi.2007.05.006.

Xing H. L., Zhang J., Yin C., 2007b. A finite element analysis of tidal deformation of the entire Earth with a discontinuous outer layer, Geophys. J. Int., 170(3): 961–970 doi:10.1111/j.1365-246X.2007.03442.x.

Xu X. (Chief Editor), 2009. Album of 5·12 Wenchun Earthquake Surface ruptures, China. Seismological Press, Beijing.

Xu X., Yang X., Yang Z., 1996. Seismogeologic disaster in urban areas and its prediction. Hydrogeol. Eng. Geol., 23(3): 32–35 (in Chinese).

Xu X., Yu G., Ma W., 2002. Evidence and methods for deter-mining the safety distance from the potential earthquake surface rupture on active fault, Seismol. Geol., 24(4): 470–483 (in Chinese).

Xu X., Wen X., Zheng R., 2003. Pattern of latest tectonic motion and its dynamics for active blocks in Sichuan-Yunnan region, China, Sci. China (Ser. D), 46(Supp): 210–226.

Xu X., Yu G., Klinger Y., Tapponnier P., Van Der Woerd J., 2006. Reevaluation of surface rupture parameters and faulting segmentation of the 2001 Kunlunshan earthquake (Mw7.8), Northern Tibetan plateau, China, J. Geophys. Res., 111: B05316, doi:10.1029/2004JB003488.

Xu X., Wen X., Chen G., Yu G., 2008a. Discovery of the Longriba fault zone in Eastern Bayan Har Block, China and its tectonic implication, Sci. China Ser. D Earth Sci., 51(9): 1209–1223.

Xu X., Wen X., Ye J., Ma B., Chen J., Zhou R., He H., Tian Q., He Y., Wang Zh., Sun Z., Feng X., Yu G., Chen L., Chen G., Yu Sh., Ran Y., Li X., Li C., An Y., 2008b. The MS8.0 Wenchuan earthquake surface ruptures and its seismogenic structure, Seismol. Geol. (in Chinese), 30(3): 597–629.

Xu X., Wen X., Yu G., Chen G., Klinger Y., Hubbard J., Shaw J., 2009a. Co-seismic reverse- and oblique-slip surface faulting generated by the 2008 Mw 7.9 Wenchuan earthquake, China, Geology, 37(6): 515–518 doi:10.1130/G25462A.1.

Xu X., Yu G., Chen G., Ran Y., Li Ch., Chen Y., Chang Ch., 2009b. Parameters of coseismic reverse- and oblique-slip surface ruptures of the 2008 Wenchuan earthquake, eastern Tibetan plateau, Acta. Geol. Sin. (English Edition), 4: 673–684.

Yu G. H., Xu X. W., Chen G. H., 2009. Relationship between the localization of earthquake surface ruptures and building damages associated with the Wenchuan 8.0 earthquake, Chin. J. Geophys. (in Chinese), 52(12): 3027–3041, doi:10.3969/j.issn.0001-5733.2009.12.012.

Yunnan Seismological Bureau, 1993. The neotectonic movement in quaternary and earthquake on the Xiaojiang fault zone (in Chinese).

Zhang P., Slemmons D. B., Mao F., 1991. Geometric pattern, rupture termination and fault segmentation of the Dixie valley—Pleasant valley active fault system, Nevada, USA, J. Struct. Geol., 13(2): 165–176.

Zhang P., Wang M., Gan W. J., 2003. Slip rates along major active faults from GPS measurements and constraints on cotemporary continental tectonics, Earth. Sci. Front., 10(special issues): 82–92 (in Chinese).

References

Zhang P., Shen Z., Wang M., Gan W., Burgmann R., Molnar P., Wang Q., Niu Z., Sun J., Wu J., Sun H., You X., 2004. Continuous deformation of the Tibetan Plateau from global positioning system data, Geology, 32: 809–812, doi: 10.1130/G20554.1.

Zhang Y., Xu L. Sh., Chen Y. T., 2009. Spatio-temporal variation of source mechanism of the 2008 Wenchuan great earthquake, Chin. J. Geophy., 52(2): 379–389.

Zhao X., Deng Q., Chen S., 1994. Tectonic geomorphology of the central segment of the Longmenshan thrust belt, western Sichuan, southern China, Seismol. Geol., 16(4): 422–428 (in Chinese).

Zhu A., Xu X., Diao G., 2008. Relocation of the Ms 8.0 Wenchuan earthquake sequence in part: preliminary seismotectonic analysis, Seismicity Geol., 30(3): 759–767.

Index

A

Acceleration, 11–13, 16–17, 179, 181
Active fault, 2, 4, 22, 26–27, 150, 179–182
Aftershock, 9, 16–19, 40, 46, 150
Altyn Tagh, 1
Anninghe, 1, 22–24, 27

B

Back-thrust pressure ridge, 39, 42, 175
Bailu Town, 40–41, 102, 107–112, 114, 124, 167, 169
Bajiao Town/Temple, 40, 45, 52–53, 102, 112, 114, 152, 178
Bayan Har, 1, 9, 11, 13–14, 28–29, 36–38
Beichuan County, 2, 6, 10, 45, 65, 67–82, 101, 149–150, 154, 156–157, 163–164, 176, 181
Beichuan-Nanba segment, 18, 39, 102
Beichuan-Yingxiu fault, 1, 40, 52, 100, 175
Bridge, 114–116, 126, 131, 140, 147, 149–151, 154, 157, 166–169

C

Chenjiaba, 80, 150
Chuan-Dian, 1, 21–23, 25–29, 36–38
Collapse, 13, 39–40, 42, 44, 47, 49–51, 55, 64, 73, 78, 106–107, 110–111, 114, 126, 136–137, 150–168, 176, 179, 181
Computation, 21, 25–26, 28
Co-seismic, 7, 12, 47, 101–102, 139, 146, 148, 168
Crustal movement, 3–4, 7
Crustal shortening, 2, 4, 44, 46, 113, 123–124

D

Daliangshan, 9, 21–22, 24, 27
Debris flow, 13, 149, 154–155, 179
Dextral pressure ridge, 39–40, 42, 44, 51, 118
Disaster, 3, 13, 149–156, 161, 167, 179–182

Double difference, 16
Dujiangyan, 45–46, 48–53, 127–128, 149, 151, 166, 170, 172–173, 175, 177–178, 180

E

Eastern Kunlun, 1
Eurasian plate, 1, 4

F

Fallen rocks, 171–172
Fault, 1–29, 36–135, 150, 153, 156, 161–165, 167–169, 175–182
Finite element, 21, 26–29, 36
Focal depth, 14, 16, 152, 181
Footwall, 71, 79, 89, 91, 114–115, 117, 124, 135, 159, 167, 176
Friction, 25, 27–29

G

Gaochuan Town, 41, 61–63, 125
Garze-Yushu, 1, 23
GPS, 3–7, 22, 28

H

Haiyuan, 1
Hanging wall, 5, 18, 39–45, 47–51, 67, 79, 83, 89, 96, 103–104, 109, 114, 116–118, 125, 131, 136, 138, 157, 164, 167–168, 176–178
Hanwang, 39–44, 102–103, 111, 118–126, 146, 151, 167, 175–176
Historical earthquake, 3, 9–10, 12, 23
Hongkou-Qingping segment, 39, 52, 102
Hongkou Town, 41, 48–52, 149, 166, 172–173, 175, 177
Hypocenter, 151–153

I

Indian plate, 1
Intraplate, 9, 39

J
Jiali fault, 1
Jianjiang River, 40, 71, 126, 130, 132, 162

K
Kara Korum fault, 1

L
Lake, 13, 72, 100, 125, 149, 152, 155, 161–162, 174
Landslide, 13, 58, 69, 101, 127–128, 149–157, 161–163, 179, 181, 182
Left-lateral, 1, 4, 9, 18, 22–24, 27, 68, 101, 126, 129–130, 132, 134–135, 147, 148
Leigu Town, 40–41, 65–69, 150
Liedu, 149
Longmenshan, 1–5, 9–11, 16–18

M
Microseismic, 10–11
Minjiang, 1, 46–48, 170, 175

N
Nanba Town, 40–41, 90–92, 150, 175
Niu Juan Gou, 151–152
Normal fault, 13–14, 39, 42, 44, 71–73, 76–77, 96–97, 101, 157, 164, 176

P
Pengguan fault, 1, 39–40, 42, 102, 124, 167
Pengxian County, 18, 176, 178
Pengzhou City, 53–55, 102–112, 126–130, 150, 167, 180
Pingtong Town, 40–42, 83–89, 150, 175
Pingwu, 6, 16, 42, 45, 83–98, 150, 175, 177
Pressure ridge, 39–48, 51–52, 55, 61, 65–69, 101, 112, 116–118, 121–122, 124, 126, 132–134, 167, 175–176

Q
Qingchuan, 18, 45, 98–100, 151, 180
Qinghai-Tibetan Plateau, 1–2, 4, 6, 21, 23, 25
Qingping Village/Town, 41, 57–58, 60–61, 150
Quaternary, 2, 24–25
Qushan Town, 6, 45, 70–75, 101, 149–150, 154–157, 161–165, 181

R
Railway, 118, 120, 157, 167, 170
Red River, 24, 27
Reverse fault, 3, 13, 17–18, 25, 39–46, 101–102, 167

Right lateral
Right lateral, 1–2, 17, 25, 39, 41–42, 45, 64, 66, 69, 71, 74, 81, 83–85, 87–92, 95, 98, 101–102, 119–124, 146–148, 175
River, 24, 27, 40, 42, 45–48, 51, 59, 63, 65, 68, 83–87, 91, 93, 96–97, 100–111, 117–126, 130–132, 136, 138, 155, 162, 168, 170, 175, 181

S
Sangzao Town, 41, 102, 125
Scarp, 6, 39–45, 47–64, 66, 68–88, 91–96, 101, 103–130, 133, 135–138, 141–143, 145, 157, 164, 167–168, 175–178
School, 110, 124, 146, 151, 154, 157, 160, 163, 166–169, 176, 178, 180
Seismic intensity, 12, 149–150, 153–154, 180–181
Seismotectonic, 1, 3
Shaba Village, 40, 74–76, 102, 117–118, 124
Shenxi Village, 40, 42, 46, 48–52
Shuiguan Village/Town, 41, 45
Sichuan Basin, 1, 4–5, 25, 28, 37–38, 151
Sichuan-Yunnan block, 1
Simple pressure ridge, 39, 42, 44, 47–48, 52, 61–62, 65–68, 121–122, 126, 175–176
Simulation, 26, 29
South China, 1, 4–5, 9, 13–14, 21, 25, 28, 37–38
Stick-slip, 25, 27
Strike-slip, 1, 9, 17–18, 22, 24, 27, 39, 45, 71, 74, 83, 85, 101–102, 126
Strong motion, 11, 13, 16–17
Surface rupture, 4–5, 9, 12, 23, 39–151, 154, 157–158, 167–168, 175–176, 179–181

T
Teaching building, 157, 160, 166, 168, 178
Tectonic, 1–3, 9, 22, 102, 182
Thrust belt, 1–2, 4, 9–11, 16–25, 29, 36–39, 42, 102, 167
Thrust scarp, 39–44, 52–53
Tianshan, 4
Tibet, 1–6, 9, 21, 23, 25, 42

U
Uplift, 1–2, 4, 6, 9, 51, 55, 63, 70, 72, 87–88, 90, 103–104, 126, 175

Index

W
Wangjiayan landslide, 156–157, 163, 181
Wenchuan-Maowen fault, 1–2, 10, 25, 39

X
Xianshuihe fault, 1, 14, 21–27, 36–38
Xiaojiang, 9, 21–22, 24–25, 27
Xiaoyudong, 18, 39–44, 126, 131–138, 148–150, 168, 176

Xichang, 24
Xuankou Town, 41, 149, 152, 154–155

Y
Yingxiu Town, 17, 45–48, 149, 152–160, 171, 175

Z
Zemuhe fault, 9, 21–22, 24, 27